The Machine in the Garden

TECHNOLOGY
AND THE PASTORAL IDEAL
IN AMERICA

Leo Marx

OXFORD
UNIVERSITY PRESS

For JANE
more than ever

OXFORD
UNIVERSITY PRESS

Oxford   New York
Athens   Auckland   Bangkok   Bogotá   Buenos Aires   Calcutta
Cape Town   Chennai   Dar es Salaam   Delhi   Florence   Hong Kong   Istanbul
Karachi   Kuala Lumpur   Madrid   Melbourne   Mexico City   Mumbai
Nairobi   Paris   São Paulo   Singapore   Taipei   Tokyo   Toronto   Warsaw

and associated companies in
Berlin   Ibadan

First published by Oxford University Press, New York, 1964
198 Madison Avenue, New York, New York 10016
First issued as an Oxford University Press paperback, 1967

Oxford is a registered trademark of Oxford University Press

Library of Congress Cataloging-in-Publication Data
Marx, Leo, 1919–
The machine in the garden : technology and the pastoral ideal in America / Leo Marx
p. cm.
Includes bibliographic references and index.
ISBN 0-19-513351-X (pbk.)
ISBN 0-19-513350-1
1. United States—Civilization. 2. Nature—Social aspects—United States.
3. Technology—Social aspects—United States. I. Title.
E169.1 .M35 2000
973—dc21   99-34697

9 8 7 6 5 4 3 2

Printed in the United States of America
on acid-free paper

Contents

ILLUSTRATIONS

THE MACHINE *in the Garden*

I *Sleepy Hollow, 1844*

I mention this peaceful spot with all possible laud; for it is in such little retired . . . valleys . . . that population, manners, and customs, remain fixed; while the great torrent of migration and improvement, which is making such incessant change in other parts of this restless country, sweeps by them unobserved. They are little nooks of still water which border a rapid stream . . .

> Washington Irving, "The Legend of Sleepy Hollow," 1820

THE pastoral ideal has been used to define the meaning of America ever since the age of discovery, and it has not yet lost its hold upon the native imagination. The reason is clear enough. The ruling motive of the good shepherd, leading figure of the classic, Virgilian mode, was to withdraw from the great world and begin a new life in a fresh, green landscape. And now here was a virgin continent! Inevitably the European mind was dazzled by the prospect. With an unspoiled hemisphere in view it seemed that mankind actually might realize what had been thought a poetic fantasy. Soon the dream of a retreat to an oasis of harmony and joy was removed from its traditional literary context. It was embodied in various utopian schemes for making America the site of a new beginning for Western society. In both forms — one literary and the other in essence political — the ideal has figured in the American view of life which is, in the widest sense, the subject of this book.

3

My purpose is to describe and evaluate the uses of the pastoral ideal in the interpretation of American experience. I shall be tracing its adaptation to the conditions of life in the New World, its emergence as a distinctively American theory of society, and its subsequent transformation under the impact of industrialism. This is not meant to be a comprehensive survey. If I were telling the story in all its significant detail, chronologically, I should have to begin at the moment the idea of America entered the mind of Europe and come down to the present — to, say, the death of Robert Frost in 1963. But I have chosen not to attempt anything so ambitious. Instead, I propose to concentrate upon selected examples, "some versions," as William Empson might put it, of American pastoralism. Nor have I confined myself to the richest, most coherent literary materials. At points I shall consider examples which have little or no intrinsic literary value. In fact, this is not, strictly speaking, a book about literature; it is about the region of culture where literature, general ideas, and certain products of the collective imagination — we may call them "cultural symbols"* — meet. To appreciate the significance and power of our American fables it is necessary to understand the interplay between the literary imagination and what happens outside literature, in the general culture. My special concern is to show how the pastoral ideal has been incorporated in a powerful metaphor of contradiction — a way of ordering meaning and value that clarifies our situation today.[1]

The notion that pastoralism remains a significant force in American life calls for an explanation. At first thought the relevance of the ancient ideal to our concerns in the second half of the twentieth century is bound to seem

---

* A "cultural symbol" is an image that conveys a special meaning (thought and feeling) to a large number of those who share the culture.

obscure. What possible bearing can the urge to idealize a simple, rural environment have upon the lives men lead in an intricately organized, urban, industrial, nuclear-armed society? The answer to this central question must start with the distinction between two kinds of pastoralism — one that is popular and sentimental, the other imaginative and complex.

1

The first, or sentimental kind is difficult to define or even to locate because it is an expression less of thought than of feeling. It is widely diffused in our culture, insinuating itself into many kinds of behavior. An obvious example is the current "flight from the city." An inchoate longing for a more "natural" environment enters into the contemptuous attitude that many Americans adopt toward urban life (with the result that we neglect our cities and desert them for the suburbs). Wherever people turn away from the hard social and technological realities this obscure sentiment is likely to be at work. We see it in our politics, in the "localism" invoked to oppose an adequate national system of education, in the power of the farm bloc in Congress, in the special economic favor shown to "farming" through government subsidies, and in state electoral systems that allow the rural population to retain a share of political power grossly out of proportion to its size. It manifests itself in our leisure-time activities, in the piety toward the out-of-doors expressed in the wilderness cult, and in our devotion to camping, hunting, fishing, picnicking, gardening, and so on. But there is no need to multiply examples; anyone who knows America today will think of many others.

Nowhere is the ill-defined feeling for "nature" more

influential than in the realm of imaginative expression. There can be little doubt that it affects the nation's taste in serious literature, reinforcing the legitimate respect enjoyed by such writers as Mark Twain, Ernest Hemingway, and Robert Frost. But on the lower plane of our collective fantasy life the power of this sentiment is even more obvious. The mass media cater to a mawkish taste for retreat into the primitive or rural felicity exemplified by TV westerns and Norman Rockwell magazine covers. Perhaps the most convincing testimony to the continuing appeal of the bucolic is supplied by advertising copywriters; a favorite strategy, validated by marketing research, assumes that Americans are most likely to buy the cigarettes, beer, and automobiles they can associate with a rustic setting.

No single motive can account for these disparate phenomena. Yet each does express something of the yearning for a simpler, more harmonious style of life, an existence "closer to nature," that is the psychic root of all pastoralism — genuine and spurious. That such desires are not peculiar to Americans goes without saying; but our experience as a nation unquestionably has invested them with peculiar intensity. The soft veil of nostalgia that hangs over our urbanized landscape is largely a vestige of the once dominant image of an undefiled, green republic, a quiet land of forests, villages, and farms dedicated to the pursuit of happiness.

In recent years several discerning, politically liberal historians of American thought have traced the gradual attenuation, in our public life, of the ideas once embodied in this cherished image. I am thinking especially of the work of Richard Hofstadter, Marvin Meyers, and Henry Nash Smith. These writers have not been concerned, to

be sure, with the relation between this body of thought
and pastoralism as a literary mode. Nor for that matter
do they often invoke the word "pastoral." But whether
they refer to "agrarianism" (the usual term), or to the
hold of "rural values" upon the national consciousness
(Hofstadter), or to the "agrarian myth" (Hofstadter), or
to the "Old Republican idyll" (Meyers), or to the "myth
of the garden" (Smith), they all seem to agree that for
some time now this tendency to idealize rural ways has
been an impediment to clarity of thought and, from their
point of view, to social progress. Anyone who shares their
assumptions is likely to find this judgment highly persuasive. They demonstrate that in public discourse, at
least, this ideal has appeared with increasing frequency
in the service of a reactionary or false ideology, thereby
helping to mask the real problems of an industrial civil-
ilization. [2]

When seen by critics of "mass culture," moreover, the
popular kind of American pastoralism assumes an equally
pernicious, if slightly different, aspect. Then it looks like
a native variant of that international form of "primitivism" to which Ortega y Gasset, among others, began
calling attention years ago. In *The Revolt of the Masses*
(1930), Ortega uses the term to describe the outlook of
a new kind of man, "a *Naturmensch* rising up in the
midst of a civilised world":

> The world is a civilised one, its inhabitant is not: he
> does not see the civilisation of the world around him, but
> he uses it as if it were a natural force. The new man
> wants his motor-car, and enjoys it, but he believes that
> it is the spontaneous fruit of an Edenic tree. In the
> depths of his soul he is unaware of the artificial, almost
> incredible, character of civilisation, and does not extend

his enthusiasm for the instruments to the principles which make them possible.

Ortega's caricature points to the shallow, not to say perverse, conception of reality inherent in our sentimental pastoralism. If his industrial *Naturmensch* bears a striking resemblance to many Americans we should not be entirely surprised. After all, what modern nation has had a history as encouraging to the illusion that its material well-being is, in Ortega's phrase, "the spontaneous fruit of an Edenic tree"? [3]

The same phenomenon aroused Sigmund Freud's interest. In his *General Introduction to Psychoanalysis* (1920), he takes up the nostalgic feeling we often attach to the unspoiled landscape as an illustration of our chronic yearning to enjoy "freedom from the grip of the external world." To Freud this impulse is the very epitome of fantasy-making:

> The creation of the mental domain of phantasy has a complete counterpart in the establishment of "reservations" and "nature-parks" in places where the inroads of agriculture, traffic, or industry threaten to change . . . the earth rapidly into something unrecognizable. The "reservation" is to maintain the old condition of things which has been regretfully sacrificed to necessity everywhere else; there everything may grow and spread as it pleases, including what is useless and even what is harmful. The mental realm of phantasy is also such a reservation reclaimed from the encroaches of the reality-principle.

Freud comes back to this subject later in *Civilization and Its Discontents* (1930). He admits that he is puzzled by what he calls the "amazing" tendency of presumably civilized men to idealize simple and often primitive conditions of life. What puzzles him most is the implication that mankind would be happier if our complex, technical order could somehow be abandoned. "How has it come about," he asks, "that so many people have adopted this strange attitude of hostility to civilization?" [4]

Freud's answer — an avowedly speculative one — is that such attitudes are the product of profound, long-standing discontent. He interprets them as signs of widespread frustration and repression. Although he assumes that every social order rests upon the denial of powerful instinctual needs, we are allowed to infer that today's advanced society may be singularly repressive. Can it be that our institutions and cultural standards are enforcing an increasingly painful, almost unbearable degree of privation of instinct? If so, this might well explain the addiction of modern man to puerile fantasies. In the light of these conjectures, the sentiments we have considered take on a pathological coloring, as if symptomatic of a collective neurosis.

Aided by the insights of Freud, Ortega, and the historians, we may begin to characterize the dominant motive back of this curious state of mind. Evidently it is generated by an urge to withdraw from civilization's growing power and complexity. What is attractive in pastoralism is the felicity represented by an image of a natural landscape, a terrain either unspoiled or, if cultivated, rural. Movement toward such a symbolic landscape also may be understood as movement away from an "artificial" world, a world identified with "art," using this word in its broadest sense to mean the disciplined habits of mind or arts developed by organized communities. In other words, this impulse gives rise to a symbolic motion away from centers of civilization toward their opposite, nature, away from sophistication toward simplicity, or, to introduce the cardinal metaphor of the literary mode, away from the city toward the country. When this impulse is unchecked, the result is a simple-minded wishfulness, a romantic perversion of thought and feeling.

If this more popular kind of pastoralism were the only kind evident in America today, we should have every reason to conclude that it is merely another of our many vehicles of escape from reality — one of those collective mental activities which can be taken seriously only for diagnostic purposes. When we turn from the general to the "high" literary culture, however, we are struck at once by the omnipresence of the same motive. One has only to consider the titles which first come to mind from the classical canon of our literature — the American books admired most nowadays — to recognize that the theme of withdrawal from society into an idealized landscape is central to a remarkably large number of them. Again and again, the imagination of our most respected writers — one thinks of Cooper, Thoreau, Melville, Faulkner, Frost, Hemingway — has been set in motion by this impulse. But while the starting point of their work and of sentimental pastoralism may be the same, the results could hardly be more different.

How shall we define that difference? The work of serious writers is different, clearly, in most of the ways that works of art differ from the flow of casual, undisciplined expression that makes up the general culture. In fact the question might easily be put aside, as it often is, simply by asserting that "literature" embodies a more sensitive and precise, a "higher," mode of perception. To do that, however, is to miss a chance of defining the complex relation between serious literature and the larger body of meanings and values, the general culture, which envelops it. An initial receptivity to the pastoral impulse is one way in which our best writers have grounded their work in the common life. But how, then, are we to explain the fact that the same impulse generates such wholly different states of mind? While in the culture at large it is the starting point for infantile wish-fulfillment dreams, a diffuse nostalgia, and a naïve, anarchic primitivism, yet it also is the source of writing that is invaluable for its power to enrich and clarify our experience. Where, then, shall we locate the point of divergence between these two modes of consciousness?

Rather than attempt to answer the question in general terms, I want to describe an event which points to an answer. Although it is an episode in the life of a writer who was to become famous, it is in other respects a typical and indeed commonplace event of the time. No doubt most of the writer's contemporaries, whether literary men or not, had similar experiences. Yet in retrospect we can see that this ordinary experience, partly because of its typicality, was one of those inconspicuous moments of discovery that has proven to be decisive in the record of our culture. What the writer discovers, though he by no means recognizes its importance, is a metaphor; he seizes upon the symbolic property or meaning in the event itself — its capacity to express much of what he thinks and feels about his situation.

2

On the morning of July 27, 1844, Nathaniel Hawthorne sat down in the woods near Concord, Massachusetts, to await (as he put it) "such little events as may happen." His purpose, so far as we can tell, was chiefly literary.

Though he had no reason to believe that anything memorable would happen, he sat there in solitude and silence
and tried to record his every impression as precisely as
possible. The whole enterprise is reminiscent of the painstaking literary exercises of his neighbor, Henry Thoreau.
Hawthorne filled eight pages of his notebook on this occasion. What he wrote is not a finished piece of work and
yet, surprisingly enough, neither is it a haphazard series
of jottings. One incident dominates the rest of his impressions. Around this "little event" a certain formal — one
might almost say dramatic — pattern takes shape. It is to
this pattern that I want to call attention.[5]

To begin, Hawthorne describes the setting, known in
the neighborhood as "Sleepy Hollow":

> . . . a shallow space scooped out among the woods, which
> surround it on all sides, it being pretty nearly circular,
> or oval, and two or three hundred yards — perhaps four
> or five hundred — in diameter. The present season, a
> thriving field of Indian corn, now in its most perfect
> growth, and tasselled out, occupies nearly half of the
> hollow; and it is like the lap of bounteous Nature, filled
> with bread stuff.

Then, in minute detail, he records what he sees and hears
close by. "Observe the pathway," he writes, "it is strewn
over with little bits of dry twigs and decayed branches,
and the sear and brown oak-leaves of last year that have
been moistened by snow and rain, and whirled about by
harsh and gentle winds, since their departed verdure . . ."
And so on. What counts here, needless to say, is not the
matter so much as the feeling behind it. Hawthorne is
using natural facts metaphorically to convey something
about a human situation. From several pages in this vein
we get an impression of a man in almost perfect repose,

idly brooding upon the minutiae of nature, and now and then permitting his imagination a brief flight. Along the path, for example, he notices that "sunshine glimmers through shadow, and shadow effaces sunshine, imaging that pleasant mood of mind where gaiety and pensiveness intermingle." For the most part, however, Hawthorne is satisfied to set down unadorned sense impressions, and especially sounds — sounds made by birds, squirrels, insects, and moving leaves.

But then, after a time, the scope of his observations widens. Another kind of sound comes through. He hears the village clock strike, a cowbell tinkle, and mowers whetting their scythes.

Without any perceptible change of mood or tone, he shifts from images of nature to images of man and society. He insists that "these sounds of labor" do not "disturb the repose of the scene" or "break our sabbath; for like a sabbath seems this place, and the more so on account of the cornfield rustling at our feet." He is describing a state of being in which there is no tension either within the self or between the self and its environment. Much of this harmonious effect is evoked by the delicate interlacing of sounds that seem to unify society, landscape, and mind. What lends most interest, however, to this sense of all-encompassing harmony and peace is a vivid contrast:

> But, hark! there is the whistle of the locomotive — the long shriek, harsh, above all other harshness, for the space of a mile cannot mollify it into harmony. It tells a story of busy men, citizens, from the hot street, who have come to spend a day in a country village, men of business; in short of all unquietness; and no wonder that it gives such a startling shriek, since it brings the noisy world into the midst of our slumbrous peace. As our

thoughts repose again, after this interruption, we find
ourselves gazing up at the leaves, and comparing their
different aspect, the beautiful diversity of green. . . .

With the train out of earshot and quiet restored, Hawthorne continues his observations. An ant colony catches
his eye. Possibly, he muses, it is the very model of the
community which the Fourierites and others are pursuing
in their stumbling way. Then, "like a malevolent genius,"
he drops a few grains of sand into the entrance of an ant
hole and obliterates it. The result is consternation among
the inhabitants, their frantic movements displaying their
"confusion of mind." How inexplicable, he writes, must
be the agency which has effected this mischief. But now
it is time for him to leave. Rising, he notices a cloud
moving across the sun; many clouds now are scattered
about the sky "like the shattered ruins of a dreamer's
Utopia. . . ." Then, in a characteristic tone of self-depre-
cation, he remarks upon the "narrow, scanty and meagre"
record of observation he has compiled during his morning in the woods. What troubles him is the discrepancy
between the shallow stream of recorded thought ("distinct
and expressed thought") and the broad tide of dim emotions, ideas, and associations that had been flowing all the
while somewhere at the back of his mind. "When we see
how little we can express," he concludes, "it is a wonder
that any man ever takes up a pen a second time."

Yet the fact is that Hawthorne has succeeded in expressing a great deal. True, there are no memorable revelations
to be got from these notes, no surprises, nothing of immediate interest from a biographical, historical, or critical
standpoint. And yet there is something arresting about
the episode: the writer sitting in his green retreat dutifully
attaching words to natural facts, trying to tap the sub-

terranean flow of thought and feeling and then, suddenly, the startling shriek of the train whistle bearing in upon him, forcing him to acknowledge the existence of a reality alien to the pastoral dream. What begins as a conventional tribute to the pleasures of withdrawal from the world — a simple pleasure fantasy — is transformed by the interruption of the machine into a far more complex state of mind.

Our sense of its evocative power is borne out by the fact that variants of the Sleepy Hollow episode have appeared everywhere in American writing since the 1840's. We recall the scene in *Walden* where Thoreau is sitting rapt in a revery and then, penetrating his woods like the scream of a hawk, the whistle of the locomotive is heard; or the eerie passage in *Moby-Dick* where Ishmael is exploring the innermost recesses of a beached whale and suddenly the image shifts and the leviathan's skeleton is a New England textile mill; or the dramatic moment in *Huckleberry Finn* when Huck and Jim are floating along peacefully and a monstrous steamboat suddenly bulges out of the night and smashes straight through their raft. More often than not in these episodes, the machine is made to appear with startling suddenness. Sometimes it abruptly enters a Happy Valley, at others a traveler suddenly comes upon it. In one of Melville's tales ("The Tartarus of Maids"), the narrator is trying to find a paper mill in the mountains; he drives his sleigh into a deep hollow between hills that rise like steep walls, and he still cannot see the place when, as he says, "suddenly a whirring, humming sound broke upon my ear. I looked, and there, like an arrested avalanche, lay the large whitewashed factory." The ominous sounds of machines, like the sound of the steamboat bearing down on the raft or of the train breaking in upon the idyll at Walden, reverberate endlessly in our literature. We hear such a sound, or see the sight which accompanies it, in *The Octopus, The Education of Henry Adams, The Great Gatsby, The Grapes of Wrath,* "The Bear" — and one could go on. Anyone familiar with American writing will recall other examples from the work of Walt Whitman, Sarah Orne Jewett, Henry James, Sherwood Anderson, Willa Cather, Eugene O'Neill, Robert Frost, Hart Crane, T. S. Eliot, John Dos Passos, Ernest Hemingway — indeed it is difficult to think of a major American writer upon whom the image of the machine's sudden appearance in the landscape has not exercised its fascination.[6]

What I am saying, in other words, is that Hawthorne's notes mark the shaping (on a microscopic scale to be sure) of a metaphoric design which recurs everywhere in our literature. They are a paradigm of the second kind of pastoralism mentioned at the outset. By looking closely at the way these notes are composed we can begin to account for the symbolic power of the "little event" in Sleepy Hollow.

3

Considered simply as a composition, as a way of ordering language to convey ideas and emotions, the first thing to notice about these casual notes is the decisive part played by the machine image. Taken by itself, what comes before we hear the train whistle scarcely arouses our interest. Descriptions of contentment seldom do. But the disturbing shriek of the locomotive changes the texture of the entire passage. Now tension replaces repose: the noise arouses a sense of dislocation, conflict, and anxiety. It is remarkable how evocative the simple device is, especially when we consider that at bottom it consists of nothing more complicated than noise clashing through harmony. This is the sensory core of the larger design, its inherent power to be revealed by its receptivity to the connotations that Hawthorne gathers about it. Like the focal point of a complicated visual pattern, this elemental, irreducible dissonance contains the whole in small.

These observations suggest the conventional character of Hawthorne's composition. For all the apparent spontaneity of his response to the event, and in spite of the novelty of the railroad in 1844 — a recent and in many ways revolutionary invention — it is striking to see how little there is here that can be called "original." One suspects indeed that if we had access to all the notebooks kept by aspiring American writers of the 1840's we would find this "little event" recorded again and again. Two years earlier, for example, one of Hawthorne's literary neighbors, Ralph Waldo Emerson, had made this entry in his journal:

> I hear the whistle of the locomotive in the woods. Wherever that music comes it has its sequel. It is the voice of the civility of the Nineteenth Century saying, "Here I am." It is interrogative: it is prophetic: and this Cassandra is believed: "Whew! Whew! Whew! How is real estate here in the swamp and wilderness? Ho for Boston! Whew! Whew! . . . I will plant a dozen houses on this pasture next moon, and a village anon. . . .

So far from being unusual, in fact, the "little event" doubtless belongs among the literary commonplaces of the age. Critics with a sociological bent often slight such a derivative aspect of the writer's response. Eager to fix his relations to his age, they look to a writer's work for direct, which is to say, spontaneous, original, unmediated reactions — as if inherited attitudes, forms, and conventions had had little or no part in shaping them. In this case, however, we have only to notice the name of the place in the woods to realize that Art, as usual, has been on the scene first. Not only has it named Sleepy Hollow, but in effect it has designed the symbolic landscape in which the industrial technology makes its appearance.[7]

The ground of Hawthorne's reaction, in other words, had been prepared by Washington Irving and Wordsworth and the "nature poets" of the previous century. In 1844, as it happens, Wordsworth wrote a sonnet protesting against the building of a railroad through the lake country. It begins: "Is then no nook of English ground secure / From rash assault? . . ." and it ends with a plea to "thou beautiful romance / Of nature" to "protest against the wrong." By placing the machine in opposition to the tranquillity and order located in the landscape, he makes it an emblem of the artificial, of the unfeeling utilitarian spirit, and of the fragmented, industrial style of life that allegedly follows from the premises of the empirical philosophy. To Wordsworth the new technology is a token of what he likes to call the "fever of the world." [8]

The pattern, moreover, can be traced back to the beginnings of industrialization. In England, as early as the 1780's, writers had been repelled by the ugliness, squalor, and suffering associated with the new factory system, and their revulsion had sharpened the taste, already strong, for images of rural felicity. We think of Blake:

> And did those feet in ancient time
>   Walk upon England's mountains green?
> And was the holy Lamb of God
>   On England's pleasant pastures seen?

> And did the Countenance Divine
> > Shine forth upon our clouded hills?
> And was Jerusalem builded here
> > Among these dark Satanic Mills?

It is evident that attitudes of this kind played an important part in quickening the massive shift in point of view which was to be called the romantic movement. Just how important they were it is difficult to say. If we regard the movement (to use Whitehead's acute phrase) as "a protest on behalf of the organic view of nature," then the contrast between the machine and the landscape would seem to embody its very essence.[9]

And yet it is misleading to think of the basic design of Hawthorne's notes as a product of modern romanticism. When we strip away the topical surface, particularly the imagery of industrialism and certain special attitudes toward visible nature, it becomes apparent that the underlying pattern is much older and more universal. Then the Sleepy Hollow motif, like a number of other conventions used by romantic writers, proves to be a modern version of an ancient literary device. It is a variation upon the contrast between two worlds, one identified with rural peace and simplicity, the other with urban power and sophistication, which has been used by writers working in the pastoral mode since the time of Virgil.

Although Theocritus is regarded as the first pastoral poet, Virgil's *Eclogues* are the true fountainhead of the pastoral strain in our literature. For one thing, in these poems Virgil (as one classical scholar puts it) "discovered" Arcadia. It is here that he created the symbolic landscape, a delicate blend of myth and reality, that was to be particularly relevant to American experience. For another, it is in the *Eclogues* that the political overtones of the pastoral situation become evident. In the background of the first eclogue, sometimes called "The Dispossessed," there was a specific action of the Roman government: the expropriation of a number of small landholders (including the poet himself) so that military veterans might be rewarded with the seized land. This display of political power no doubt intensified Virgil's feeling for the land as a symbolic repository of value; at the same time it compelled him to acknowledge the implacable character of the forces threatening the established order. Both responses are accommodated by the thematic structure of Virgil's poem; let us consider it in greater detail.[10]

The poem takes the form of a dialogue between two shepherds. Tityrus, like Virgil, has successfully petitioned for the return of his land. At the outset he is happily playing upon his pipe when Melibœus, who has been evicted, comes by with his herd. Here are the opening lines as translated by E. V. Rieu:

> Tityrus, while you lie there at ease under the awning of a spreading beech and practise country songs on a light shepherd's pipe, I have to bid good-bye to the home fields and the ploughlands that I love. Exile for me, Tityrus — and you lie sprawling in the shade, teaching the woods to echo back the charms of Amaryllis.

Tityrus answers with praise of the patron in Rome to whom he owes his liberty and his "happy leisure." The man gave his word, says Tityrus, "and my cattle browse at large, while I myself can play the tunes I fancy on my rustic flute." In reply, Melibœus disclaims any feeling of jealousy. "*My* only feeling is amazement — with every farm in the whole countryside in such a state of chaos. Look at myself, unfit for the road, yet forced to drive my

goats on this unending trek." He points to one animal
he can "hardly drag" along. "Just now," he explains,
". . . she bore two kids — I had been counting on them
— and had to leave the poor things on the naked flints."
He berates himself for not anticipating "this disaster."

The first eclogue certainly represents more than a simple wish-image of bucolic pleasure. No sooner does Virgil
sketch in the ideal landscape than he discloses an alien
world encroaching from without. Melibœus represents this
other world. Through his lines we are made aware that
the immediate setting, with its tender feeling and contentment, is an oasis. Beyond the green hollow the countryside is in a state of chaos. The very principle of natural
fecundity is threatened (he has been forced to abandon
his newborn kids). What is out there, from the reader's
point of view, is a world like the one he inhabits; it contains great cities like Rome, organized power, authority,
restraint, suffering, and disorder. We are made to feel that
the rural myth is threatened by an incursion of history.
The state of mind of Melibœus — we should call it alienation nowadays — brings a countervailing force to bear
upon the pastoral ideal. Divested of his land, he faces the
prospect of unending anxiety, deprivation, and struggle:

> . . . the rest of us are off; some to foregather with the
> Africans and share their thirst; others to Scythia, and
> out to where the Oxus rolls the chalk along; others to
> join the Britons, cut off as they are by the whole width
> of the world. Ah, will the day come, after many years,
> when I shall see a place that I can call my home . . . ?

The whole thrust of the poem is toward a restoration of
the harmony established in the opening lines. Lying at
ease under the beech, playing his pipe, Tityrus embodies
the pastoral ideal. Here, incidentally, the distinction be-

tween the pastoral and primitive ideals may be clarified. Both seem to originate in a recoil from the pain and responsibility of life in a complex civilization — the familiar impulse to withdraw from the city, locus of power and politics, into nature. The difference is that the primitivist hero keeps going, as it were, so that eventually he locates value as far as possible, in space or time or both, from organized society; the shepherd, on the other hand, seeks a resolution of the conflict between the opposed worlds of nature and art. Since he often is the poet in disguise — Tityrus represents Virgil himself — he has a stake in both worlds. In the first eclogue nothing makes the mediating character of the pastoral ideal so clear as the spatial symbolism in which it is expressed. The good place is a lovely green hollow. To arrive at this haven it is necessary to move away from Rome in the direction of nature. But the centrifugal motion stops far short of unimproved, raw nature. "Happy old man!" the unfortunate Melibœus says to his friend: "So your land will still be yours. And it's enough for you, even though the bare rock and marshland with its mud and reeds encroach on all your pastures. Your pregnant ewes will never be upset by unaccustomed fodder; no harm will come to them. . . ."

This ideal pasture has two vulnerable borders: one separates it from Rome, the other from the encroaching marshland. It is a place where Tityrus is spared the deprivations and anxieties associated with both the city and the wilderness. Although he is free of the repressions entailed by a complex civilization, he is not prey to the violent uncertainties of nature. His mind is cultivated and his instincts are gratified. Living in an oasis of rural pleasure, he enjoys the best of both worlds — the sophisticated order of art and the simple spontaneity of nature. In a few lines

Virgil quickly itemizes the solid satisfactions of the pastoral retreat: peace, leisure, and economic sufficiency. The key to all of these felicities is the harmonious relation between Tityrus and the natural environment. It is a serene partnership. In the pastoral economy nature supplies most of the herdsman's needs and, even better, nature does virtually all of the work. A similar accommodation with the idealized landscape is the basis for the herdsman's less tangible satisfactions: the woods "echo back" the notes of his pipe. It is as if the consciousness of the musician shared a principle of order with the landscape and, indeed, the external universe. The echo, a recurrent device in pastoral, is another metaphor of reciprocity. It evokes that sense of relatedness between man and not-man which lends a metaphysical aspect to the mode; it is a hint of the quasi-religious experience to be developed in the romantic pastoralism of Wordsworth, Emerson, and Thoreau. Hence the pastoral ideal is an embodiment of what Lovejoy calls "semi-primitivism"; it is located in a middle ground somewhere "between," yet in a transcendent relation to, the opposing forces of civilization and nature.[11]

What is most impressive, when we read the first eclogue with Hawthorne's notes in mind, is the similarity of the root conflict and of the over-all pattern of thought and emotion. By his presence alone Melibœus reveals the inadequacy of the Arcadian situation as an image of human experience. His lines convey the intervention of reality; they are a check against our susceptibility to idyllic fantasies. In 1844 Hawthorne assigns a similar function to the machine. Like Virgil's unfortunate herdsman, the sound of the locomotive "brings the noisy world into the midst of . . . slumbrous peace." Although the railroad is a recent invention (the first American railroad had begun operations in 1829), many of the associations it is made to carry are more or less timeless features of *the* world, that is to say, the great world as it traditionally had been conceived in literature from the Old Testament to the poetry of Wordsworth. The train stands for a more sophisticated, complex style of life than the one represented by Sleepy Hollow; the passengers are "busy men, citizens, from the hot street. . . ." The harsh noise evokes an image of intense, overheated, restless striving — a life of "all unquietness" like that associated with great cities as far back as the story of the tower of Babel. The central device of Hawthorne's notes is to expose the pastoral ideal to the pressure of change — to an encroaching world of power and complexity or, in a word, to history. It is a modern variant of the design of Virgil's poem.

4

What, to be more precise, then, is a pastoral design? And how is this particular design related to the two kinds of pastoralism introduced at the outset?

By *design* I refer to the larger structure of thought and feeling of which the *ideal* is a part. The distinction is a vital one. Much of the obscurity that surrounds the subject stems from the fact that we use the same word to refer to a wish-image of happiness and to literary compositions in their entirety — pastoral dreams and pastoral poems. Then, too, we sometimes confuse matters even more by taking the word completely out of its literary context to describe our experience of the real world. We say of a pleasing stretch of country that it is a "pastoral scene," or that it gives us a "pastoral feeling." But our reactions to literature seldom are that simple. (The confusion arises, as it so often does, in crossing the borderland between life and literature.) Most literary works called pastorals — at least those substantial enough to retain our interest — do not finally permit us to come away with anything like the simple, affirmative attitude we adopt toward pleasing rural scenery. In one way or another, if only by virtue of the unmistakable sophistication with which they are composed, these works manage to qualify, or call into question, or bring irony to bear against the illusion of peace and harmony in a green pasture. And it is this fact that will enable us, finally, to get at the difference between the complex and sentimental kinds of pastoralism.

In addition to the ideal, then, the pastoral design in question (it is one among many) embraces some token of a larger, more complicated order of experience.* Whether represented by the plight of a dispossessed herdsman or by the sound of a locomotive in the woods, this feature of the design brings a world which is more "real" into juxtaposition with an idyllic vision. It may be called the *counterforce*. Admittedly, the portentous, melodramatic connotations of this term make it inappropriate for the discussion of many bland, pre-industrial versions of pastoral. (Among the more effective of the traditional counters to the pastoral dream have been certain stylized tokens of mortality. We should understand that the counterforce may impinge upon the pastoral landscape either from the side bordering upon intractable nature or the side facing

---

* The scope of the design need not be an entire work; it may be confined to a scene or episode (a "pastoral interlude") within a poem, drama, or novel which is not, strictly speaking, a pastoral. I regard those works as pastorals whose controlling theme is a variant of the conflict between art and nature — nature being represented by an idealized image of landscape.

advanced civilization. During the seventeenth century, Poussin and other landscape painters introduced the image of a speaking death's-head into the most delicate pictorial idylls. To make the meaning of this *memento mori* inescapable they sometimes inserted the printed motto, *Et in Arcadia Ego,* meaning "I [Death] also am in Arcadia.") * Nevertheless, the term *counterforce* is applicable to a good deal of modern American writing. The anti-pastoral forces at work in our literature seem indeed to become increasingly violent as we approach our own time. For it is industrialization, represented by images of machine technology, that provides the counterforce in the American archetype of the pastoral design.[12]

Since Jefferson's time the forces of industrialism have been the chief threat to the bucolic image of America. The tension between the two systems of value had the greatest literary impact in the period between 1840 and 1860, when the nation reached that decisive stage in its economic development which W. W. Rostow calls the "take-off." In his study of the more or less universal stages of industrial growth, Rostow defines the take-off as the "great watershed in the life of modern societies" when the old blocks and resistances to steady development are overcome and the forces of economic progress "expand and come to

---

* The device figures what Erwin Panofsky calls a discrepancy "between the supernatural perfection of an imaginary environment and the natural limitations of human life as it is." It is typical of the unceasing metamorphosis of the pastoral mode that sometime around 1630 the original meaning of this device was lost to view. In an illuminating essay, Panofsky has shown that at this time the motto was reinterpreted so that the words, instead of being attributed to Death itself, were taken as the sentiments of another shepherd. Thus what had been intended as a dramatic encounter with death was replaced by a relatively sentimental and tranquillizing idea, in consonance with the main drift of the age.

dominate the society." In America, according to Rostow, the take-off began about 1844 — the year of the Sleepy Hollow episode — just at the time our first significant literary generation was coming to maturity. Much of the singular quality of this era is conveyed by the trope of the interrupted idyll. The locomotive, associated with fire, smoke, speed, iron, and noise, is the leading symbol of the new industrial power. It appears in the woods, suddenly shattering the harmony of the green hollow, like a presentiment of history bearing down on the American asylum. The noise of the train, as Hawthorne describes it, is a cause of alienation in the root sense of the word: it makes inaudible the pleasing sounds to which he had been attending, and so it estranges him from the immediate source of meaning and value in Sleepy Hollow. In truth, the "little event" is a miniature of a great — in many ways the greatest — event in our history.[13]

That Hawthorne was fully aware of the symbolic properties of the railroad is beyond question. Only the year before he had published "The Celestial Railroad," a wonderfully compact satire on the prevailing faith in progress. In the popular culture of the period the railroad was a favorite emblem of progress — not merely technological progress, but the overall progress of the race. Hawthorne's sketch turns upon the idea of the new machine as a vehicle for an illusory voyage of salvation whose darkest meanings are reserved for readers of Bunyan. Like the hero of *The Pilgrim's Progress,* the American pilgrim thinks he is on his way to the Heavenly City. As it turns out, however, the same road can lead to hell, the partly concealed point being that the American protagonist is not a Christian at all; he has much more in common with the other traveler in Bunyan's Calvinist allegory, Ignorance.

Nevertheless, it would be wrong to suppose that the primary subject of the Sleepy Hollow notes is the transition from an agrarian to an industrial society. Plausible on its face, such a reading confuses literary ends and means. The whole tenor of the notes indicates that Hawthorne is not interested in directing attention from himself to what is happening "out there" in the great world of political and institutional change. Nor can it be said, incidentally, that any of the works that embody variants of the motif are, in the usual sense, *about* the great transformation. The point may seem a niggling one, but it is crucial if we are to define the precise relation between literature and that flow of unique, irreversible events called history. Although Hawthorne's account includes an element of representation — he draws upon actual objects and events — his chief concern is the landscape of the psyche. The inner, not the outer world, is what interests him most as he sits there in the woods, attempting to connect words and sense perceptions. His aim, as he says, is to represent the broad tide of dim emotions, ideas, and images coursing through his mind. When he seizes upon the auditory image of the train it is because it serves this purpose.

The primary subject of the Sleepy Hollow notes, then, is the contrast between two conditions of consciousness. Until he hears the train's whistle Hawthorne enjoys a serenity close to euphoria. The lay of the land represents a singular insulation from disturbance, and so enhances the feeling of security and repose. The hollow is a virtual cocoon of freedom from anxiety, guilt, and conflict — a shrine of the pleasure principle. To describe the situation in the language of Freud, particularly when we have only one example in view, no doubt seems farfetched. But the

striking fact is that again and again our writers have introduced the same overtones, depicting the machine as
invading the peace of an enclosed space, a world set apart,
or an area somehow made to evoke a feeling of encircled
felicity. The setting may be an island, or a hut beside a
pond, or a raft floating down a river, or a secluded valley
in the mountains, or a clearing between impenetrable
walls of forest, or the beached skeleton of a whale — but
whatever the specific details, certain general features of
the pattern recur too often to be fortuitous. Most important is the sense of the machine as a sudden, shocking
intruder upon a fantasy of idyllic satisfaction. It invariably
is associated with crude, masculine aggressiveness in contrast with the tender, feminine, and submissive attitudes
traditionally attached to the landscape.

But there is no need, actually, to choose between the
public and private, political and psychic, meanings of this
event. Even in these offhand notes, Hawthorne's first concern — he is, after all, a writer of fiction — is the emotional power of his material, and that power unquestionably is heightened by the larger, political implications of
the machine image. Emerson makes the point this way:
the serious artist, he says, "must employ the symbols in
use in his day and nation to convey his enlarged sense to
his fellow-men." The ideas and emotions linked to the
fact of industrialization provide Hawthorne with just such
an enlargement of meaning. Their function is like that
of the secondary subject, or "vehicle," of a grand metaphor.
To say this is not to imply that the topical significance of
the machine is "extrinsic" to the literary process, or that
it may be treated as a merely illustrative appendage. As
with any well-chosen figure, the subsidiary subject is an
integral part of the metaphor. Thought and feeling flow

both ways. The radical change in the character of society
and the sharp swing between two states of feeling, between
an Arcadian vision and an anxious awareness of reality,
are closely related: they illuminate each other. All of
which is another way of accounting for the symbolic
power of the motif: it brings the political and the psychic
dissonance associated with the onset of industrialism into
a single pattern of meaning. Once generated, of course,
that dissonance demands to be resolved.[14]

At the end of Virgil's poem the resolution is effected
by a series of homely images. Tityrus invites Melibœus to
postpone his journey into exile. "Yet surely," he says, "you
could sleep here as my guest for this one night, with green
leaves for your bed?" This symbolic gesture may be interpreted as an offer of a "momentary stay against confusion" — Robert Frost's way of defining the emotional end-
product of a poem. "It begins in delight," says Frost, "and
ends in wisdom . . . in a clarification of life — not necessarily a great clarification, such as sects or cults are
founded on, but in a momentary stay against confusion."
To objectify this state of equilibrium Virgil closes the
gap between hope and fear. In the last lines of the poem
Tityrus blends the two emotions in a picture of the landscape at twilight:

> I have got ripe apples, and some mealy chestnuts and a
> good supply of cheese. See over there — the rooftops of
> the farms are already putting up their evening smoke
> and shadows of the mountain crests are falling farther
> out.[15]

So ends the first eclogue. As far as the narrative is concerned, nothing has been solved. The poem offers no hint
of a "way out" for Melibœus or those who inhabit the
ravaged countryside. All that he gets for solace is one

night's postponement of his exile — one night of comfort
and companionship. Yet this twilight mood, a blend of
sadness and repose, succeeds aesthetically. It is a virtual
resolution. Like the middle landscape, or the ritual marriage at the end of a pastoral romance, this consolatory
prospect figuratively joins what had been apart. At the
end of the Sleepy Hollow notes, similarly, the train moves
off and a sad tranquillity comes over Hawthorne. Although he manages to regain some of his earlier sense of
peace, the encroaching forces of history have compelled
him to recognize its evanescence. Just as Virgil ends with
the image of falling shadows, so Hawthorne ends with the
thought of a "dreamer's Utopia" in ruins. In each case the
conflict aroused by the counterforce is mitigated. These
highly stylized resolutions are effective partly because the
writers succeed in maintaining an unruffled, contemplative, Augustan tone. This tone, characteristic of Virgilian
pastoral, is a way of saying that the episode belongs to a
timeless, recurrent pattern of human affairs. It falls easily
into a conventional design because it has occurred often
before.

But the fact is that nothing quite like the event announced by the train in the woods had occurred before.
A sense of history as an unpredictable, irreversible sequence of unique events makes itself felt even in Hawthorne's notes. In spite of the resemblance between the
train and the archetypal city of Western literature, the
"little event" creates an unprecedented situation. For in
the stock contrast between city and country each had been
assumed to occupy a more or less fixed location in space:
the country here, the city there. But in 1844 the sound of
a train in the Concord woods implies a radical change
in the conventional pattern. Now the great world is in-

vading the land, transforming the sensory texture of rural life — the way it looks and sounds — and threatening, in fact, to impose a new and more complete dominion over it. (Compare plates 1 and 2.) True, it may be said that agents of urban power had been ravaging the country-side throughout recorded history. After they had with-drawn, however, the character of rural life had remained essentially unchanged. But here the case is different: the distinctive attribute of the new order is its tech-nological power, a power that does not remain con-fined to the traditional boundaries of the city. It is a centrifugal force that threatens to break down, once and for all, the conventional contrast between these two styles of life. The Sleepy Hollow episode prefigures the emer-gence, after 1844, of a new, distinctively American, post-romantic, industrial version of the pastoral design. And the feelings aroused by this later design will have the ef-fect of widening the gap, already great, between the pastoralism of sentiment and the pastoralism of mind.

At the outset I introduced the "little event" of 1844 to mark the shaping of a metaphor, or metaphoric design, which appears again and again in modern American writ-ing. With Virgil's poem in view, however, we can see that the episode does not represent the beginning so much as the decisive turning point of a long story. It would be more accurate, then, to say that Hawthorne, in seizing upon the image of the railroad as counterforce, is re-shap-ing a conventional design to meet the singular conditions of life in nineteenth-century America. To understand his response to the machine we must appreciate the intensity of his feeling for its opposite, the landscape. The same may be said of many American writers. Their heightened sensitivity to the onset of the new industrial power can only be explained by the hold upon their minds of the pastoral ideal, not as conceived by Virgil, but as it had been adapted, since the age of discovery, to New World circumstances. What those circumstances were, and how they influenced the development of native pastoralism, both sentimental and complex, is the burden of what follows.

If any man shall accuse these reports of partiall falshood,
supposing them to be but Utopian, and legendarie fables,
because he cannot conceive, that plentie and famine, a temperate climate, and distempered bodies, felicities, and miseries can be reconciled together, let him now reade with
judgement, but let him not judge before he hath read.

     *A True Declaration of the estate of the Colonie
     in Virginia . . . , London, 1610*

SOME of the connections between *The Tempest*
and America are well known. We know, for one thing,
that Shakespeare wrote the play three or possibly four
years after the first permanent colony had been established at Jamestown in 1607. At the time all of England
was in a state of excitement about events across the Atlantic. Of course, the play is not in any literal sense about
America; although Shakespeare is nowhere explicit about
the location of the "uninhabited island," so far as he allows us to guess it lies somewhere in the Mediterranean
off the coast of Africa. For the dramatist's purpose it might
be anywhere. Nevertheless, it is almost certain that Shakespeare had in mind the reports of a recent voyage to the
New World. In 1609 the *Sea Adventure*, one of a fleet of
ships headed for Virginia, was caught in a violent storm
and separated from the rest. Eventually it ran aground in
the Bermudas and all aboard got safely to shore. Several

people wrote accounts of the episode, and unmistakable echoes of at least two of them may be heard in *The Tempest,* particularly in the storm scene. At one point, moreover, Ariel refers to having fetched dew from the "still-vex'd Bermoothes." Though all of these facts are well known and reasonably well established, they do not in themselves suggest a particularly significant relation between the play and America. They indicate only that Shakespeare was aware of what his countrymen were doing in the Western hemisphere.[1]

But when, in addition to the external facts, we consider the action of *The Tempest,* a more illuminating connection with America comes into view. The play, after all, focuses upon a highly civilized European who finds himself living in a prehistoric wilderness. Prospero's situation is in many ways the typical situation of voyagers in newly discovered lands. I am thinking of the remote setting, the strong sense of place and its hold on the mind, the hero's struggle with raw nature on the one hand and the corruption within his own civilization on the other, and, finally, his impulse to effect a general reconciliation between the forces of civilization and nature. Of course, this is by no means a uniquely American situation. The conflict between art and nature is a universal theme, and it has been a special concern of writers working in the pastoral tradition from the time of Theocritus and Virgil. Besides, the subject has a long foreground in Shakespeare's own work — witness *A Midsummer Night's Dream, As You Like It,* and *The Winter's Tale.* Nevertheless, the theme is one of which American experience affords a singularly vivid instance: an unspoiled landscape suddenly invaded by advance parties of a dynamic, literate, and purposeful civilization. It would be difficult to imagine a more dramatic

coming together of civilization and nature. In fact, Shakespeare's theme is inherent in the contradictory images of
the American landscape that we find in Elizabethan travel
reports, including those which he seems to have read before writing *The Tempest*.

1

Most Elizabethan ideas of America were invested in visual
images of a virgin land. What most fascinated Englishmen was the absence of anything like European society;
here was a landscape untouched by history — nature unmixed with art. The new continent looked, or so they
thought, the way the world might have been supposed to
look before the beginning of civilization. Of course the
Indians also were a source of fascination. But their simple
ways merely confirmed the identification of the New
World with primal nature. They fit perfectly into the picture of America as a mere landscape, remote and unspoiled, and a possible setting for a pastoral retreat. But
this does not mean that Shakespeare's contemporaries
agreed about the character or the promise of the new land.
Quite the contrary. Europeans never had agreed about the
nature of nature; nor did they now agree about America.
The old conflict in their deepest feelings about the
physical universe was imparted to descriptions of the terrain. Elizabethan travel reports embody sharply contrasting images of the American landscape.

At one extreme, among the more popular conceptions,
we find the picture of America as paradise regained. According to his account of a voyage to Virginia in 1584,
Captain Arthur Barlowe was not yet in sight of the coast
when he got a vivid impression that a lovely garden lay

ahead. We "found shole water," he writes, "wher we smelt so sweet, and so strong a smel, as if we had bene in the midst of some delicate garden abounding with all kinde of odoriferous flowers. . . ." Barlowe, captain of a bark dispatched by Sir Walter Raleigh, goes on to describe Virginia in what was to become a cardinal image of America: an immense garden of "incredible abundance." The idea of America as a garden is the controlling metaphor of his entire report. He describes the place where the men first put ashore as

> . . . so full of grapes, as the very beating and surge of the Sea overflowed them, of which we found such plentie, as well there as in all places else, both on the sand and on the greene soile of the hils . . . that I thinke in all the world the like abundance is not to be found: and my selfe having seene those parts of Europe that most abound, find such difference as were incredible to be written.

Every detail reinforces the master image: Virginia is a land of plenty; the soil is "the most plentifull, sweete, fruitfull, and wholsome of all the worlde"; the virgin forest is not at all like the "barren and fruitles" woods of eastern Europe, but is full of the "highest and reddest Cedars of the world." One day Barlowe watches an Indian catching fish so rapidly that in half an hour he fills his canoe to the sinking point. Here Virginia stands not only for abundance, but for the general superiority of a simple, primitive style of life. Geography controls culture: the natives are "most gentle, loving and faithfull, voide of all guile and treason, and such as live after the maner of the golden age." [2]

The familiar picture of America as a site for a new golden age was a commonplace of Elizabethan travel litera-

ture, and there are many reasons for its popularity. For
one thing, of course, the device made for effective propa-
ganda in support of colonization. Projects like those of
Raleigh required political backing, capital, and colonists.
Even in the sixteenth century the American countryside
was the object of something like a calculated real estate
promotion. Besides, fashionable tendencies in the arts
helped to popularize the image of a new earthly paradise.
During the Renaissance, when landscape painting emerged
as a separate genre, painters discovered — or rather, as
Kenneth Clark puts it, rediscovered — the garden. The
ancient image of an enchanted garden gave the first serious
painters of landscape their most workable organizing motif.
To think about landscape at all in this period, therefore,
was to call forth a vision of benign and ordered nature.
And a similar concern makes itself felt in Elizabethan
literature. Pastoral poetry in English has never in any
other period enjoyed the vogue it had then. The exploration of North America coincided with the publication of
Spenser's Virgilian poem, *The Shepheards Calendar* (1579)
and Sidney's *Arcadia* (1590), to name only two of the
more famous Elizabethan pastorals. It is impossible to
separate the taste for pastoral and the excitement, felt
throughout Europe, about the New World. We think of
the well-known golden age passage in *Don Quixote* and
Michael Drayton's *Poems Lyrick and Pastoral,* both of
which appeared about 1605. Drayton's volume included
"To the Virginian Voyage," with its obvious debt to Captain Barlowe's report. He praises "VIRGINIA / Earth's
only paradise," where

> . . . Nature hath in store
> Fowl, venison, and fish,

And the fruitfull'st soil,
Without your toil,
Three harvests more,
All greater than your wish.

. . . .

To whom the golden age
Still Nature's laws doth give,
  Nor other cares that tend
  But them to defend
From winter's rage,
That long there doth not live.

As in Barlowe's report, the new land smells as sweet to the approaching voyager as the most fragrant garden:

When as the luscious smell
Of that delicious land,
  Above the sea's that flow's,
  The clear wind throws,
Your hearts to swell
Approaching the dear strand.

The age was fascinated by the idea that the New World was or might become Arcadia, and we hardly need to itemize the similarities between the "gentle, loving, and faithfull" Indians of Virginia and the shepherds of pastoral. In Elizabethan writing the distinction between primitive and pastoral styles of life is often blurred, and devices first used by Theocritus and Virgil appear in many descriptions of the new continent.[3]

Although fashionable, the image of America as a garden was no mere rhetorical commonplace. It expressed one of the deepest and most persistent of human motives. When Elizabethan voyagers used this device they were drawing upon utopian aspirations that Europeans always had cherished, and that had given rise, long before the discovery of America, to a whole series of idealized, imaginary worlds. Besides the golden age and Arcadia, we are reminded of Elysium, Atlantis, and enchanted gardens, Eden and Tirnanogue and the fragrant bower where the Hesperides stood watch over the golden apples. Centuries of longing and revery had been invested in the conception. What is more, the association of America with idyllic places was destined to outlive Elizabethan fashions by at least two and a half centuries. It was not until late in the nineteenth century that this way of thinking about the New World lost its grip upon the imagination of Europe and America. As for the ancillary notion of the new continent as a land of plenty, that, as we all know, is now stronger than ever. Today some historians stress what the sixteenth-century voyager called "incredible abundance" as perhaps the most important single distinguishing characteristic of American life. In our time, to be sure, the idea is less closely associated with the landscape than with science and technology.[4]

Elizabethans, however, did not always fancy that they were seeing Arcadia when they gazed at the coast of North America. Given a less inviting terrain, a bad voyage, a violent storm, hostile Indians, or, most important, different presuppositions about the universe, America might be made to seem the very opposite of a bountiful garden. Travelers then resorted to another conventional metaphor of landscape depiction. In 1609, for example, when William Strachey's vessel reached the New World, it was caught, as he puts it, in "a dreadfull storme and hideous . . . which swelling, and roaring as it were by fits, . . . at length did beate all light from heaven; which like an hell of darkenesse turned blacke upon us. . . ." After the ship was beached, Strachey and his company realized that they were on one of the "dangerous and dreaded . . . Ilands of the Bermuda." His report is one that Shakespeare probably was thinking about when he wrote *The Tempest*. In this "hideous wilderness" image of landscape, the New World is a place of hellish darkness; it arouses the fear of malevolent forces in the cosmos, and of the cannibalistic and bestial traits of man. It is associated with the wild men of medieval legend.[5]

No doubt the best-known example of this reaction appears in William Bradford's account of an event that occurred shortly after the Bermuda wreck. When the *Mayflower* stood off Cape Cod in September 1620, Bradford (as he later recalled) looked across the water at what seemed to him a "hidious and desolate wilderness, full of wild beasts and willd men." Between the pilgrims and their new home, he saw only "deangerous shoulds and roring breakers." So far from seeming an earthly paradise, the landscape struck Bradford as menacing and repellent.

> Nether could they, as it were, goe up to the tope of Pisgah, to vew from this willdernes a more goodly cuntrie to feed their hops; for which way soever they turnd their eys (save upward to the heavens) they could have litle solace or content in respecte of any outward objects. For summer being done, all things stand upon them with a wetherbeaten face; and the whole countrie, full of woods and thickets, represented a wild and savage heiw. If they looked behind them, ther was the mighty ocean which they had passed, and was now as a maine barr and goulfe to separate them from all the civill parts of the world.

This grim sight provoked one of the first of what has been an interminable series of melancholy inventories of the

desirable — not to say indispensable — items of civilization absent from the raw continent. His people, said
Bradford, had "no freinds to wellcome them, nor inns to
entertaine or refresh their weatherbeaten bodys, no houses
or much less townes to repaire too, to seeke for succoure."
Instead of abundance and joy, Bradford saw deprivation
and suffering in American nature.[6]

Here, then, is a conception of the New World that is
radically opposed to the garden. On the spectrum of Elizabethan images of America the hideous wilderness appears
at one end and the garden at the other. The two views are
traditionally associated with quite different ideas of man's
basic relation to his environment. We might call them
ecological images. Each is a kind of root metaphor, a
poetic idea displaying the essence of a system of value.
Ralph Waldo Emerson had some such concept in mind
when he observed, in *English Traits,* that the views of
nature held by any people seem to "determine all their
institutions." In other words, each image embodies a quite
distinct notion of America's destiny — a distinct social
ideal.[7]

To depict the new land as a lovely garden is to celebrate
an ideal of immediate, joyous fulfillment. It must be admitted, however, that the word "immediate" conceals a
crucial ambiguity. How immediate? we may well ask. At
times the garden is used to represent the sufficiency of
nature in its original state. Then it conveys an impulse-
centered, anarchic, or primitivistic view of life. But elsewhere the garden stands for a state of cultivation, hence
a less exalted estimate of nature's beneficence. Although
important, the line between the two is not sharp. Both
the wild and the cultivated versions of the garden image
embody something of that timeless impulse to cut loose

from the constraints of a complex society. In Elizabethan travel literature the image typically carries a certain sense of revulsion — quickened no doubt by the discovery of new lands — against the deprivation and suffering that had for so long been accepted as an unavoidable basis for civilization. To depict America as a garden is to express aspirations still considered utopian — aspirations, that is, toward abundance, leisure, freedom, and a greater harmony of existence.

To describe America as a hideous wilderness, however, is to envisage it as another field for the exercise of power. This violent image expresses a need to mobilize energy, postpone immediate pleasures, and rehearse the perils and purposes of the community. Life in a garden is relaxed, quiet, and sweet, like the life of Virgil's Tityrus, but survival in a howling desert demands action, the unceasing manipulation and mastery of the forces of nature, including, of course, human nature. Colonies established in the desert require aggressive, intellectual, controlled, and well-disciplined people. It is hardly surprising that the New England Puritans favored the hideous wilderness image of the American landscape.[8]

What is most revealing about these contrasting ideas of landscape is not, needless to say, their relative accuracy in picturing the actual topography. They are not representational images. America was neither Eden nor a howling desert. These are poetic metaphors, imaginative constructions which heighten meaning far beyond the limits of fact. And yet, like all effective metaphors, each had a basis in fact. In a sense, America was *both* Eden and a howling desert; the actual conditions of life in the New World did lend plausibility to both images. The infinite resources of the virgin land really did make credible, in minds long habituated to the notion of unavoidable scarcity, the ancient dream of an abundant and harmonious life for all. Yet, at the same time, the savages, the limitless spaces, and the violent climate of the country did threaten to engulf the new civilization. In the reports of voyagers there was evidence to support either view, and during the age of Elizabeth many Englishmen seized upon one or the other as representing the truth about America and her prospects.

But there were others who recognized the contradiction and attempted to understand or at least to express it. Sylvester Jourdain, who wrote a report on the Bermuda wreck of 1609, observes that the islands were widely considered "a most prodigious and inchanted place, affoording nothing but gusts, stormes, and foule weather; which made every Navigator and Mariner to avoide them . . . as they would shunne the Devill himselfe. . . ." It was all the more surprising, therefore, when the castaways discovered that the climate was "so temperate and the Country so aboundantly fruitful of all fit necessaries" that they were able to live in comfort for nine months. Experience soon led them to reconsider the legendary horror of the place. Jourdain (one of the writers with whom Shakespeare apparently was familiar) finally puts it this way: "whereas it [Bermuda] hath beene, and is still accounted, the most dangerous infortunate, and most forlorne place of the world, it is in truth the richest, healthfullest, and pleasing land, (the quantity and bignesse thereof considered) and meerely naturall, as ever man set foote upon." [9]

William Strachey, in his report, also confronts the ambiguity of nature in the New World. As already mentioned, he had been impressed by the legendary hideousness of the islands:

> . . . they be so terrible to all that ever touched on them,
> and such tempests, thunders, and other fearefull objects
> are seene and heard about them, that they be called
> commonly, The Devils Ilands, and are feared and avoyded
> of all sea travellers alive, above any other place in the
> world.

Then in the very next sentence Strachey acknowledges the
contrary evidence: "Yet it pleased our mercifull God, to
make even this hideous and hated place, both the place of
our safetie, and meanes of our deliverance." By invoking
Providence he can admit the attractions of the islands
without revising the standard opinion of Bermuda as a
hideous wilderness. There were a number of devices for
coping with these contradictory ideas about America. One
writer, anxious to correct the dismal reports about life at
Jamestown during the early years, attacks the problem
head on. Having begun with a stock and no doubt transparently propagandistic celebration of Virginia's abundance, and aware at the same time that the actual calami-
ties were well known, he interjects the direct appeal to
the reader's credulity that I quote at the head of this
chapter. For him the problem is to persuade readers to
accept an image of America in which "felicities and
miseries can be reconciled together. . . ." [10]

But if some Elizabethan travelers discovered that the
stock images of America embraced a contradiction, few
had the wit to see what mysteries it veiled. Few recognized
that a most striking fact about the New World was its
baffling hospitality to radically opposed interpretations.
If America seemed to promise everything that men always had wanted, it also threatened to obliterate much of
what they already had achieved. The paradox was to be
a cardinal subject of our national literature, and begin-

ning in the nineteenth century our best writers were able
to develop the theme in all its complexity. Not that the
conflict was in any sense peculiar to American experience.
It had always been at the heart of pastoral; but the discovery of the New World invested it with new relevance,
with fresh symbols. Nothing demonstrates this fact more
clearly than the play Shakespeare wrote in the hour
colonization began.

2

ADRIAN.   Though this island seem to be desert,—
SEBASTIAN. Ha, ha, ha!

. . . .

ADRIAN.   Uninhabitable, and almost inaccessible, —
SEBASTIAN. Yet, —
ADRIAN.   Yet, —
ANTONIO.   He could not miss't.
ADRIAN.   It must needs be of subtle, tender and deli-
cate temperance.
ANTONIO.   Temperance was a delicate wench.
SEBASTIAN. Ay, and a subtle; as he most learnedly
deliver'd.
ADRIAN.   The air breathes upon us here most sweetly.
SEBASTIAN. As if it had lungs, and rotten ones.
ANTONIO.   Or as 'twere perfum'd by a fen.
GONZALO.   Here is everything advantageous to life.
ANTONIO.   True; save means to live.
SEBASTIAN. Of that there's none, or little.
GONZALO.   How lush and lusty the grass looks! how
green! [11]

This exchange takes place when the court party first
examines the island after the wreck; it is a comic version
of the effort to reconcile conflicting attitudes toward the

New World. But, for all the jesting, a genuine sense of the terrain — its palpable presence — comes through. The actuality of the landscape, hence the close juxtaposition of fact and fancy, is a distinguishing mark of pastoral set in the New World. To be sure, a remote and unspoiled landscape had long been a feature of the mode. But the usual setting of pastorals had been a never-never land. The writer did not pretend that it was an actual place, and the reader was not expected to take it as one. (In Europe, for one thing, it was difficult to credit the existence of a site that was both ideal and unoccupied.) In the age of discovery, however, a note of topographical realism entered pastoral. Writers were increasingly tempted to set the action in a terrain that resembled, if not a real place, then the wish-colored image of a real place. Even when the connection is not made explicit, as in *The Tempest,* we surely feel the imaginative impact of an actual New World.

This sense of discovery accounts in part for the close affinity between the travel literature and Elizabethan pastoral. Many voyagers resort to pastoral conventions in writing their reports. What gives them a special stamp, indeed, is the close juxtaposition of the conventional and the novel; artificial devices of rhetoric are used to report fresh, striking, geographical facts. The combination heightens the sense of awe at the presence of a virgin continent. Although we may recognize Virgil's shepherds in Barlowe's Indians, who are gentle and loving after the manner of the golden age, the fact remains that the place is Virginia. Virginia does exist. Can it be, then, that the old, old dream suddenly has come true? The question, even when unexpressed, makes itself felt in much of the

early writing about America, and it lends fresh conviction
and immediacy to the pastoral impulse. We feel it in
Gonzalo's exclamation: "How lush and lusty the grass
looks! how green!" The greenness of grass is hardly a
novelty, yet the conventionality of the image lends credibility to the episode. The effect comes from the strength
of feeling attached to so homely a fact. Could anything
but green grass, actually before his eyes, produce so lyrical
a banality? "The singing of the little birds," Christopher
Columbus had written in 1492 of a West Indian island,
"is such that it seems that one would never desire to
depart hence." [12]

But what kind of place is Shakespeare's "uninhabited
island"? Like Arcadia or Virginia, it is remote and unspoiled, and at first thought we are likely to remember
it as a kind of natural paradise. The play leaves us with
a memory of an enchanted isle where life is easy and the
scenery a delight. Or if this is not the whole truth (after
all, what about Caliban?), it surely is clear that the island
is the sort that can be expected to arouse utopian fantasies
in the minds of Europeans. Had not the voyages of Columbus and Cabot inspired Sir Thomas More's *Utopia* in
1516? As soon as the storm is over and the castaways re-
assemble on the beach we hear the discussion just cited. It
is Gonzalo, the "honest old Councellor," who is most responsive to the promise of the green island. Soon he is
half-seriously ruminating about the kingdom he would
set up if he had "plantation of this isle."

> GONZALO.   I' th' commonwealth I would by contraries
>   Execute all things; for no kind of traffic
>   Would I admit; no name of magistrate;
>   Letters should not be known; riches, poverty,
>   And use of service, none; contract, succession,

> Bourn, bound of land, tilth, vineyard, none;
> No use of metal, corn, or wine, or oil;
> No occupation; all men idle, all;
> And women too, but innocent and pure:
> No sovereignty; —
> SEBASTIAN.   Yet he would be King on 't.
> ANTONIO.   The latter end of his commonwealth forgets
>   the beginning.
> GONZALO.   All things in common Nature should produce
> Without sweat or endeavour: treason, felony,
> Sword, pike, knife, gun or need of any engine,
> Would I not have; but Nature should bring forth,
> Of it own kind, all foison, all abundance,
> To feed my innocent people.[13]

How seriously shall we take the old man's vision of the island's possibilities? Most scholars say that Shakespeare merely is mocking the whole idea. We know that he borrowed it, and indeed some of the actual words, from Montaigne's essay on cannibals — one of the fountainheads of modern primitivism. We also know that the speech belongs to a tradition that goes back to the ancient Greek idea of man's original state. Gonzalo admits as much a moment later: "I would with such perfection govern, sir, / T' excel the Golden Age." But it is one thing to identify Shakespeare's source and quite another to know what to make of this famous passage. We sense immediately that the world of Gonzalo's imagination, for all his "merry fooling," is in many ways similar to the "real" world of the play, the enchanted island that Prospero rules. Yet it is impossible to miss the skepticism that Shakespeare places, like a frame, around the old man's speech. We feel it in the force that he lends to the interruptions of the insolent courtiers, Sebastian and Antonio.

These two are nothing if not worldly wise, and they quickly and shrewdly detect the veiled egoism in Gonzalo's conception of society. Antonio puts it with impressive economy: "The latter end of his commonwealth forgets the beginning." In one line Shakespeare condenses a treatise on a fallacy that nullifies most primitivist-anarchist programs. The scornful courtier sees that Gonzalo is befuddled about the uses of power, and that he proposes to exercise absolute power in order to set up a polity dedicated to the abolition of power. Gonzalo means to imply, of course, that here nature is so benign that power is not necessary. Hence he would begin his imaginary regime by dispensing with government, learning, technology — even agricultural technology. He is dreaming of what a nineteenth-century utopian might call the "withering away of the state." To see how far Shakespeare is from sharing these sentiments, we have only to compare the "beginning" of Gonzalo's utopia with the "beginning" of Prospero's actual island "commonwealth." [14]

We get our first impression of the setting for Prospero's regime from the violent storm with which *The Tempest* begins. In the opening scene Shakespeare creates a kind of tempestuous no-man's-land between civilization and this other, new world. Later, it is true, we learn that the storm had been contrived by Prospero himself, but then we also recognize that he is forcing his old enemies to re-enact his own passage from civilization into nature. When, twelve years before, he and Miranda also had survived a tempest in a "rotten carcass of a butt," they had been at the mercy of the elements. Nor does Prospero ever forget that the elements were merciless. His enemies, he tells Miranda, launched them upon the water:

> To cry to th' sea that roar'd to us; to sigh
> To th' winds, whose pity, sighing back again,
> Did us but loving wrong.

Like the castaways in William Strachey's report of the Bermuda wreck, Prospero and Miranda were saved only by "Providence divine." [15]

The opening scene dramatizes the precariousness of civilization when exposed to the full fury of nature. In seventy spare lines we are given what Strachey, describing the Bermuda tempest, calls a "dreadfull storme and hideous . . . swelling, and roaring as it were by fits. . . ." Against a background of thunder and lightning an eggshell ship founders on a furious ocean. " 'We split, we split!' — 'Farewell, my wife and children!' " Disorder in society follows close upon disorder in nature. In the emergency the lowly boatswain, who gives orders to noble courtiers, justifies his disregard of social degree by pointing to nature itself: "What cares these roarers for the name of King?" So far as he is concerned, what counts in the crisis is seamanship, technical skill, the ability to resist and repress primal forces. But Gonzalo, who even then fails to appreciate the need for power, thinks the seaman impudent: ". . . remember," he warns him, "whom thou hast aboard." Whereupon the boatswain, emboldened by danger, invites the nobleman to prove his authority over the tempest — "command these elements" — or, in effect, keep still. [16]

Shakespeare leaves no doubt about the difference between the original state of nature as Gonzalo imagines it and as it actually exists in the world of *The Tempest*. The audience at the play enters this world through a howling storm, and so experiences something of what Prospero and Miranda had found when they arrived on the island.

We are likely to forget that the place then was under the sway of evil forces. Its ruler was Caliban's mother, the malevolent "blue-ey'd hag," Sycorax, and twelve years later Caliban still thinks of the island as his rightful possession. "This island's mine, by Sycorax my mother, / Which thou tak'st from me." Before she died the "damn'd witch" had imprisoned Ariel in a cloven pine, and to win his co-operation Prospero reminds Ariel what the place had once been like:

> Then was this island —
> Save for the son that she did litter here,
> A freckled whelp hag-born — not honour'd with
> A human shape.[17]

In its "original" state, that is, before Prospero's arrival, this new world had been a howling desert, where the profane ruled. In the first scene Shakespeare uses every possible device to stress the violent, menacing power of nature. Above all, he makes the storm scene a scene apart. The rest of the action is colored by fantasy, but the storm is depicted in spare naturalistic tones. We are invited to imagine a real ship in a real tempest. The contrast between this scene and the rest of the play is underscored by the absence of Prospero. This is the only time when we are unaware of the controlling power of his magic. As soon as the scene is over we learn of his art, and from then on nothing is "natural," all is touched by enchantment. The rest of the action, until Prospero announces in the last act that the "charm dissolves apace," is a kind of dream. What is more, the dramatic structure of *The Tempest* sets this opening scene off from the rest. Critics frequently note how close Shakespeare comes to observing the classical unities in the play — how (except for the storm scene) all the action takes place on the island, and how (except for the storm scene) it is all neatly packed into four hours. But in unifying the rest of the play Shakespeare isolates and thereby accentuates the force of the storm itself. It lends its name to the entire action. How much simpler the stagecraft would have been had the play begun with a scene on the beach after the wreck (the facts about the storm might easily have been conveyed by dialogue), but how much imaginative force would have been lost! To carry its full dramatic weight the storm must be dramatized. In that way Shakespeare projects an image of menacing nature, and of the turmoil that Prospero had survived, into dramatic time. The opening scene represents the furies, without and within the self, that civilized man must endure to gain a new life. As an ironic underpinning for Gonzalo's sentimental idea of nature, nothing could be more effective than the howling storm.[18]

Nor is the menacing character of nature confined to the opening episode. Within the encircling storm, to be sure, there is a lush, green island; but in depicting it Shakespeare does not allow us to forget the hideous wild. All through the first of the comic episodes, when Trinculo and Stephano meet Caliban, we hear thunder rolling in the distance. "Alas, the storm is come again!" says Trinculo as he crawls under the monster's gaberdine. If we sometimes lose sight of unruly nature during the play, it is largely because Prospero's art had done so much of its work by the time the action begins. In twelve years he has changed the island from a howling desert into what seems an idyllic land of ease, peace, and plenty. With his magic he has eliminated or controlled many unpleasant, ugly features of primal nature. And yet enough remain to dispel the primitivistic daydream that Gonzalo speaks.[19]

Apart from the initial storm, the most vivid reminder of the island's past is Caliban. It is from Caliban that Prospero learns an invaluable lesson about unimproved nature. At first his feelings about primitive man, like Gonzalo's, had been those of the more optimistic Elizabethan voyagers. He had responded to Caliban as Gonzalo responds to the "Shapes" that Prospero conjures up:

> If in Naples
> I should report this now, would they believe me?
> If I should say, I saw such islanders, —
> For, certes, these are people of the island, —
> Who, though they are of monstrous shape, yet, note,
> Their manners are more gentle, kind, than of
> Our human generation you shall find
> Many, nay, almost any.

As he reminds Caliban, Prospero once had thought it possible to nurture and redeem him:

> I have us'd thee,
> Filth as thou art, with human care; and lodg'd thee
> In mine own cell, till thou didst seek to violate
> The honour of my child.

Like her father, Miranda also had pitied Caliban before the abortive attack. She had taught him human speech, but now she refers to him as: "Abhorred slave, / Which any print of goodness wilt not take, / Being capable of all ill!" It is true that Prospero has Caliban in his power when the action begins, but the creature's threatening presence reminds us throughout that the dark, hostile forces exhibited by the storm are still active. We are not in Eden; Caliban must be made to work. He keeps us in mind of the unremitting vigilance and the repression of instinct necessary to the felicity Prospero and Miranda enjoy.[20]

But finally it is Prospero himself who most clearly defines the nature of nature, and man's relation to it, in the exotic setting of *The Tempest*. Before his exile he had led an almost exclusively passive and contemplative life, "rapt in secret studies." If there is still something of the medieval hermit about Prospero as we see him, he now recognizes what was wrong with his earlier life. By "neglecting worldly ends," he had helped the conspiracy that took his throne. Since then he has been forced to exercise as well as study power. His survival and his triumph rest upon art, a white magic akin to science and technology. As readers of *The Golden Bough* or the work of Malinowski know, there are close affinities between magic and modern science, particularly in their tacit views of man's necessary posture in the face of physical nature. Both presuppose our ability and our need to master the non-human through activity of mind. The aim of Prospero's magic, as his relations with Ariel and Caliban show, is to keep the elements of air, earth, fire, and water at work in the service of his island community. He does not share Gonzalo's faith in what "Nature should produce / Without sweat or endeavour." [21]

Each of these attitudes toward nature accords with a distinct idea of history. Like all primitivist programs, Gonzalo's plantation speech in effect repudiates calculated human effort, the trained intellect, and, for that matter, the idea of civilization itself. It denies the value of history. It says that man was happiest in the beginning — in the golden age — and that the record of human activity is a record of decline. But Prospero's personal history exemplifies a contrary view. At first, after his banishment, the elements controlled his fate, but gradually, by use of reason and art, he won dominion over nature. As his name suggests, he is a kind of meliorist. (The names of several characters in *The Tempest* have overtones of symbolic significance. Caliban, for example, is formed from the letters of "cannibal.") Prospero is derived from the Latin, *prosperare,* to cause to succeed. Although a reclusive scholar in Europe, here in this new world Prospero is more like a social engineer. Given the setting, it may not be too farfetched to see his behavior as prophetic of the deliberate and sometimes utopian manipulation of social forms that would tempt Europeans in a virgin land. His sense of the plastic character of human behavior is largely unstated, but it is made graphic by his impressive role as "designer" of the drama itself. It is he, after all, who stands above the level of ordinary mortals and contrives the ritual of initiation that finally achieves a change of state.

As the shaping spirit of the play, Prospero directs the movement toward redemption, not by renouncing power, but by exercising it to the full. His control is based upon hard work, study, and scholarly self-discipline. We are constantly reminded of his studious habits, and even Caliban recognizes that his power stems from the written word. "Remember/," he warns his fellow conspirators, "First to possess his books; for without them / He's but a sot, as I am, nor hath not / One spirit to command. . . ." Until the final scene we are kept mindful of Prospero's nagging sense of responsibility, and his devotion to the reasoned use of power. What he feels toward the external forces of nature, moreover, has its counterpart in what he feels about passion, his own included. When Ariel reports the anguish of his enemies, Prospero has to master his vengeful impulse:

> Though with their high wrongs I am struck to th' quick,
> Yet with my nobler reason 'gainst my fury
> Do I take part. . . .

As man must control the animals, so must intellect dominate passion. Though Prospero is a model of self-mastery, taut, humorless, and awesome, he does not neglect the emotional and sensual aspect of man. This humane balance appears in his self-effacing and magnanimous encouragement of Miranda's suitor. At the same time he characteristically insists upon a controlled and chaste engagement. Prospero fulfills Hamlet's ideal: the man who is not passion's slave.[22]

Prospero's experience represents a denial of the idea, expressed by Gonzalo, that we should emulate the spontaneous, uncalculated ways of mindless nature. In the beginning, according to the skeletal fable of *The Tempest,* there was more of chaos than of order — a fearful storm, not a delightful garden. To reach the island, an uninhabited site for a new beginning, Europeans must risk annihilation. Prospero's success, finally, is the result not of submission to nature, but of action — of change that stems from intellect. On this island power makes love possible, and so both may be said to rest upon the book of magical lore — that is, upon the fusion of mind and object best exemplified by art. In Prospero's triumph Shakespeare affirms an intellectual and humanistic ideal of high civilization.

3

But having said all this, we still must cope with what now seems a paradox: our initial impression that in *The Tempest* Shakespeare glorifies nature, the island landscape, and the rusticity of Prospero's little community. To conclude that Prospero's triumph is a triumph of art over nature does not square with our full experience of the

play. We cannot forget that here redemption is made possible by a journey away from Europe into the wilderness.
In Milan the hero's art was anything but triumphant; on
the island it helps to create a kind of natural paradise.
Here is a life of unexampled bounty, serenity, freedom,
and delight. It is an idealized rural style of life, in striking contrast to what we may infer about life back in
Milan. The simple comforts, the dreamlike remoteness
from the stress of the great world, the lyrical immersion
in the immediate sensations of nature — we identify all
of these boons with the unspoiled landscape. Above all,
we think of the setting as conducive to a sensuous inti-
macy between man and not-man, like the echoes in Virgil's
first eclogue, that nourishes the spirit as well as the body.
But how then are we to reconcile the triumph of Prospero's
intellectual ideal with a celebration of the movement away
from advanced society toward nature?

Some critics, it should be said, deny that *The Tempest*
embraces a positive view of physical nature or, to be more
concrete, of the island setting. But perhaps that is because
Shakespeare, like many writers working in the pastoral
tradition, relies upon auditory images to carry his strongest
affirmation of man's rootedness in nature. The visible
landscape is relatively insignificant in *The Tempest,* but
to conclude that the characters are unaffected by the setting is to ignore what Shakespeare does with music. The
uninhabited island is a land of musical enchantment, and
from beginning to end all kinds of engaging sounds play
upon our ears. We hear music being sung and played
and talked about; we hear the music of the poetry itself;
and, what is most revealing, music often serves as the
audible expression of external nature's influence upon
mind. After the wreck, the first survivor we meet is Ferdi-

nand, who is following sounds that seem to flow from the very terrain itself:

> Where should this music be? i' th' air or th' earth?
> It sounds no more: and, sure, it waits upon
> Some god o' th' island. Sitting on a bank,
> Weeping again the King my father's wrack,
> This music crept by me upon the waters,
> Allaying both their fury and my passion
> With its sweet air. . . .

That this music actually is made by Ariel is a point I will come back to. Now we need only observe that the music *seems* to emanate from the air and earth and water. It is one with the landscape, and its effect upon Ferdinand is similar to the effect of the setting upon several of the others. We think of Adrian: "The air breathes upon us here most sweetly," or Gonzalo: "How lush and lusty the grass looks! how green!" [23]

Music puts at rest the fury of the storm. The opposition between music and the tempest is symbolic of the deepest conflict in the play. In fact, one critic (G. Wilson Knight) maintains that these polar images are the key to the dominant unifying theme of the entire Shakespearean canon. Here, in any event, there is no mistaking the power of music to allay the forces of disorder. Even Caliban, as readers often note, responds to the melodious atmosphere:

> . . . the isle is full of noises,
> Sounds and sweet airs, that give delight, and hurt not.
> Sometimes a thousand twangling instruments
> Will hum about mine ears; and sometimes voices . . .

Caliban's bestiality, the equivalent within human nature of the untamed elements without, is partly offset by his singular, heavy-footed grace of language. (If the language was given him by Miranda, the feeling is his own.) And it is heard, significantly, in homage to the island's bountiful landscape:

> I'll show thee the best springs; I'll pluck thee berries;
> I'll fish for thee, and get thee wood enough.
>
> .   .   .   .
>
> I prithee, let me bring thee where crabs grow;
> And I with my long nails will dig thee pig-nuts;
> Show thee a jay's nest, and instruct thee how
> To snare the nimble marmoset; I'll bring thee
> To clustering filberts, and sometimes I'll get thee
> Young scamels from the rock. Wilt thou go with me?

Monstrous though he is, there is in Caliban a vein of crude tenderness that makes itself felt largely as susceptibility to music and landscape. As a result he is more appealing than several of the deceitful, corrupt, and besotted men of Europe washed up on this beach. In *The Tempest* music is a measure of the beauty, order, and proportion of the physical universe.[24]

But how, then, can we interpret the story of Prospero's exile, the experience of "reality" in *The Tempest,* as a total repudiation of Gonzalo's utopian fantasy? If the plantation speech rests upon the idea of a beneficent order running through nature, so in a way does the Pythagorean musical motif. So does Prospero's delicate masque. For that matter, the entire fable unquestionably affirms the impulse of civilized man to renew himself by immersion in the simple, spontaneous instinctual life. Witness Ariel, the spirit of air who helps Prospero recognize what is truly human. Once an aloof and haughty scholar, Prospero learns on the island the compassion that finally allows him to restore his enemies to themselves. At the moment, already mentioned, when he controls his impulse toward revenge, it is Ariel who gives him his cue:

ARIEL. . . . if you now beheld them, your affections
Would become tender.

PROS.   Dost thou think so, spirit?

ARIEL.   Mine would, sir, were I human.

PROS.   And mine shall.
Hast thou, which art but air, a touch, a feeling
Of their afflictions, and shall not myself,
One of their kind, that relish all as sharply
Passion as they, be kindlier mov'd than thou art?
Though with their high wrongs I am struck to th' quick,
Yet with my nobler reason 'gainst my fury
Do I take part . . .

Or, as another token of nature's promise, witness Prospero's success in bringing his daughter up to perfect womanhood in this remote place. Here, far from the sophistication of Europe, he actually does create a brave new world, even if Miranda ingenuously (and how ironically!) confuses it with an old ignoble one.[25]

But why, then, if he does not permit us to take Gonzalo and his dreamland seriously, does Shakespeare expect us to feel differently about Prospero and his New World Arcadia? The answer may be found by comparing the plantation speech and the lovely masque of Act IV.

There are good reasons for thinking of the masque — Ferdinand calls it a "most majestic vision" — as the counterpart, for Prospero, of Gonzalo's vision of the perfect plantation. Each in its way is a tribute to the principle of natural fecundity. Each depicts an ideal land of abundance and joy. Each has its inception in an image of landscape. At the beginning of the masque, Iris (accompanied by soft music) sets the stage by addressing Ceres:

Ceres, most bounteous lady, thy rich leas
Of wheat, rye, barley, vetches, oats, and pease;

> *Thy turfy mountains, where live nibbling sheep,*
> *And flat meads thatch'd with stover, them to keep;*

Later the goddess of agriculture responds by blessing the betrothed couple in these words:

> *Earth's increase, foison plenty,*
> *Barns and garners never empty;*
> *Vines with clust'ring bunches growing;*
> *Plants with goodly burthen bowing;*
> *Spring come to you at the farthest*
> *In the very end of harvest!*
> *Scarcity and want shall shun you;*
> *Ceres' blessing so is on you.*

Like the imaginary plantation, this is in effect another openhearted utopian vision. Indeed, if we listen closely to Ceres we can hear a distinct echo (notice the repetition of "foison") of Gonzalo's speech; in his plantation, he had said:

> . . . Nature should bring forth,
> Of it own kind, all foison, all abundance,
> To feed my innocent people.

The affinity between the two visions is heightened by the warm relation between the two men. When he first sees his old friend, Prospero cries, "Holy Gonzalo, honourable man," and again, a moment later, "O good Gonzalo, / My true preserver." [26]

And yet the differences between the two finally are more revealing than the similarities. Prospero's masque is a dream, as is the plantation speech, but it is a dream far more consistent with what *The Tempest* tells us of reality. Here the landscape, so far from representing Eden or the original state of nature, is an idealized version of old England — a countryside that men have acted upon for a long time. It is the traditional domain of Ceres, that is, of *agri - culture* (in the Latin: fields plus culture), an amalgam of landscape and art. If the land now looks like a magnificent garden, there is no reason to doubt that it once was a hideous wilderness. This paradise is a product of history in a future partly designed by men. (Perhaps there is more than a graceful compliment in Ceres' lines: "Spring come to you at the farthest / In the very end of harvest!") The successful blending of art and nature colors all the final affirmations of *The Tempest*. The music of the island is not made of the native woodnotes wild we might expect to hear in Gonzalo's perfect plantation. Rather it is a product of collaboration between Prospero and Ariel. Miranda, by the same token, combines the qualities of natural simplicity with breeding and education. Her presence requires us to take seriously the prospect of utopia. She has the gift of wonder. (Her name is derived from the Latin, *mirari*, to wonder, or *mirus*, wonderful.) Hence her response when she first sees the court party:

> O wonder!
> How many goodly creatures are there here!
> How beauteous mankind is! O brave new world,
> That has such people in't! [27]

The tone of the masque carries over into the final moments of the play, which are filled with the joy that follows the success of controlled human effort. It is Prospero's effort that is being rewarded, and the tone differs markedly from the tone of Gonzalo's vision, with its yearning for a soft, passive, and indolent style of life. All of Prospero's behavior, including the masque, suggests that half-formed, indistinct idea of history as a record of human

improvement, or progress, that was incipient in Renaissance thought generally. His commitment to art for the
melioration of life reminds us of Francis Bacon, who at
this time not only wrote an important chapter in the history of the idea of progress, but was convinced that the
superiority of the present to the past could be explained
by specific innovations in the practical arts. It is a testimony to the power of the poetic imagination that this
hermit, with his magical incantations, also can seem a
prophet of the emergent faith in progress.[28]

But the difference between Gonzalo and Prospero is not
that one accepts and the other rejects the idea of utopia.
Like his gentle friend, Prospero also delights in dreams
of the good life. But he does not lose sight of the line
that separates dream and reality. Shakespeare dramatizes
the point most tellingly when, without warning, Prospero
stops the masque. Before that, to enhance the effect, we
are shown the powerful grip that these poetic images take
upon the mind. As he watches the little play, Ferdinand
exclaims: "Let me live here ever;/ So rare a wonder'd
father and a wise / Makes this place Paradise." Ferdinand
is carried away, and so (he admits in an aside) is Prospero.
That is when he *starts suddenly, and speaks; after which,
to a strange, hollow and confused noise. . . ."* the actors
vanish. Prospero "had forgot that foul conspiracy/ Of the
beast Caliban and his confederates. . . ." Twelve years
before, in Milan, he had remained "rapt" in the life of
the mind while an attack was being mounted against him.
He is not about to repeat that mistake. He has learned
that the urgencies of power take precedence, at least in
order of time, over visionary pleasures. Yet there is nothing in what he says or does to belittle the masque. Like
Gonzalo, Prospero values utopian visions. But his dream

is closer, within the world of *The Tempest,* to "reality" or the possible. In fact, the masque may be taken as an oblique statement of the principles for which he now goes forth to suppress Caliban and redeem his European enemies.[29]

The difference between the masque and the plantation speech, finally, is the difference between a pastoral and a primitive ideal. For Prospero the center of value is located in the traditional landscape of Ceres. He stands on a middle ground, a terrain of mediation between nature and art, feeling and intellect. Any inclination that he might have had to trust in primal nature (as indicated by his original attitude toward Caliban) now has been checked by his experience of the hideous wilderness, by what he knows of the storm and of his own aggressive impulses and, of course, by what he has come to accept as the truth about Caliban. Even in the closing moments of the play, when he forgives his former enemies, he says nothing to suggest a change in his estimate of Caliban as "A devil, a born devil, on whose nature / Nurture can never stick. . . ." His recognition of inherent and perhaps irremediable aggressiveness in man saves Prospero's utopian bent from sentimentality. But neither does he go to the opposite extreme. That he has reservations about the cultivated man, about power, intellect, and art, is implied by his final act of renunciation. In the end he abjures the potent art that distinguishes him from ordinary men. As if distrusting the uses of power, he vows to bury his staff and drown his book. The action of *The Tempest* has the effect of mitigating the duality that first sets it off. It is a comedy in praise neither of nature nor of civilization, but of a proper balance between them.[30]

4

I began by saying that there are revealing connections, beyond the facts about Shakespeare's sources, between *The Tempest* and America. The first connection is genetic. It links the theme of the play to the contradiction within the Elizabethan idea of the New World. Is the virgin land best described as a garden or a hideous wilderness? In *The Tempest,* as in the travel reports, we can find apparent confirmation of both images. The island, like America, could be Eden *or* a hellish desert. As the action progresses, the weight of its implication swings like a pendulum between the poles of nature and civilization. The opening movement, when Milan comes to the wilderness, describes the widest arc: a vast distance between the howling storm, in which art counts for everything, and Gonzalo's primitivist vision of Eden regained, where nature counts for everything. But as Prospero's power makes itself felt, the arc becomes shorter. His aim is reconciliation, and as he masters the situation, the pendulum slows down; we move from storm to calm, from discord toward harmony. In the end we are permitted a glimpse, but only a glimpse, of the possibility of man's earthly transfiguration.

But what justification are we given for so utopian a vision? Leaving aside the force of the convention itself (this is, after all, a comedy), we may ask whether the ending is a tribute to Prospero's art or to the musical order, latent in nature, represented by Ariel? The answer implied by the last act is that neither art nor nature can be the basis for hope. So far as the play affirms the pastoral ideal of harmony, it draws upon Shakespeare's sense of an underlying unity that binds consciousness to the energy and order manifest in unconscious nature. If a resolution of the pastoral conflict is conceivable, Shakespeare implies, it is because art itself is a product of nature. In *The Winter's Tale,* a play closely related to *The Tempest* both in spirit and time of composition, he had made his most searching statement of the theme. Perdita and Polixenes are discussing the relative merit of wild and man-bred flowers. In Perdita's garden there are no cultivated flowers:

>   . . . of that kind
> Our rustic garden's barren; and I care not
> To get slips of them.
> POLIXENES.   Wherefore, gentle maiden,
> Do you neglect them?
> PERDITA.   For I have heard it said
> There is an art which in their piedness shares
> With great creating Nature.
> POLIXENES.   Say there be;
> Yet Nature is made better by no mean
> But Nature makes that mean; so, over that art
> Which you say adds to Nature, is an art
> That Nature makes. You see, sweet maid, we marry
> A gentler scion to the wildest stock,
> And make conceive a bark of baser kind
> By bud of nobler race. This is an art
> Which does mend Nature, change it rather, but
> The art itself is Nature.

The context, it is generally conceded, lends Shakespeare's support to Polixenes' view of the matter: the artificial is but a special, human category of the natural. Mind and nature are in essence one. Nature is all. This conviction underlies the seriousness with which Shakespeare, in *The Tempest,* treats the pastoral ideal.[31]

But what is there about *The Tempest* that imparts a

rare and compelling credibility, even profundity, to what for so long had seemed an incredible ideal? No single answer will do, but it is clear that certain realities of the age of discovery lend this ancient fantasy its astonishing force and depth. A universal awareness of vast, remote, and unspoiled continents had renewed the plausibility of the pastoral dream, now projected into the future. Although most earlier versions of pastoral had been set in never-never lands, and although *The Tempest* contains only one allusion to the actual New World, its setting is not wholly fanciful. We begin with a commonplace event of the age: a ship caught in a storm and beached on an uninhabited island. It is like an Elizabethan news report. Beginning with this episode enables Shakespeare to avoid the artificiality, the initial wrenching away from the world of ordinary experience, that we expect of pastoral. To be sure, we move swiftly from the tempest into a world of dreamlike enchantment, but, nevertheless, we move there from an event that has the plausibility of the actual. On this island we encounter no courtiers masquerading as shepherds; these are castaways, survivors of a wreck, and their situation keeps us in mind of the real world beyond. And though the body of the play is closer in texture to myth than to reality, the action carries us from the actual to the mythical and back again. Like the verse, with its vibrant sense of place, the action reflects the poet's awareness of an unspoiled terrain — a new world that really exists. The imaginative authority of the fable arises from the seriousness and wonder with which Shakespeare is able to depict a highly civilized man testing his powers in a green and desolate land.

In addition to the genetic connection between *The Tempest* and America, there is another that can only be called prophetic. By this I mean that the play, in its over-all design, prefigures the design of the classic American fables, and especially the idea of a redemptive journey away from society in the direction of nature. As in *Walden, Moby-Dick,* or *Huckleberry Finn,* the journey begins with a renunciation. The hero gives up his place in society and withdraws toward nature. But in *The Tempest,* as in the best of American pastoral, the moral significance of this move is ambiguous. The island, after all, is the home of Caliban, who embodies the untrammeled wild-ness or cannibalism at the heart of nature. And that fact also proves to be a comment upon human nature, as Prospero indicates at the end: "Two of these fellows you / Must know and own," he says, "this thing of darkness I / Acknowledge mine." In *The Tempest* the island is not an ideal place, any more than the woods are in *Walden* or the sea in *Moby-Dick* or the river in *Huckleberry Finn.* And yet, in a world that contains corrupt Milan, the island does offer hope. Precisely because it is untainted by civilization, man's true home in history, it offers the chance of a temporary return to first things. Here, as in a dream, the superfluities and defenses of everyday life are stripped away, and men regain contact with essentials. In the wilderness only essentials count. America, Emerson will say, is a land without history, hence a land "where man asks questions for which man was made." [32]

What finally enables us to take the idea of a successful "return to nature" seriously is its temporariness. It is a journey into the desert and back again — "a momentary stay against confusion." On the island Prospero regains access to sources of vitality and truth. This we must grant even if we deny that the island, representing external nature, provides anything more than a setting for the

renewal that Prospero achieves through an effort of mind
and spirit. What happens during his exile is what may
happen to us in any of our departures from routine waking
consciousness. It is what may happen in sleep, especially
in dreams (the action of *The Tempest* is a kind of dream),
in the act of love, perhaps even in death where the race
renews itself if only in making room for the newborn. It
is not necessary to commit the pathetic fallacy in order to
accept the restorative power of this movement, literal or
symbolic, away from the city toward nature. The contrast
between "city" and "country" in the pastoral design makes
perfect sense as an analogue of psychic experience. It
implies that we can remain human, which is to say, fully
integrated beings, only when we follow some such course,
back and forth, between our social and natural (animal)
selves. In Milan Prospero's sense of value had become dis-
torted; he had succumbed to the tyranny of his scholarly
ambitions. But on the island he rediscovers what it means
to be a father, to have senses and a passional self. What
is ascribed to "nature" in the design may plausibly be
understood as the vitality of unconscious or preconscious
experience. The partnership between these two realms of
being is a means as well as an end in Shakespeare's fable.
The resolution is accomplished by Prospero working with
Ariel. Even as the final unravelling draws near, Prospero
requires Ariel's help — "some heavenly music" to lead the
court party to the climax. In the closing scene several of
the "lost" Europeans find each other and themselves:

>   . . . in one voyage
> Did Claribel her husband find at Tunis,
> And Ferdinand, her brother, found a wife
> Where he himself was lost, Prospero his dukedom
> In a poor isle, and all of us ourselves,
> When no man was his own.[33]

And so to Naples and then to Milan. The play fosters no illusion that a permanent retreat from the city is possible or desirable. But the temporary exile, or psychic renewal, may also be understood in political terms. If the city is corrupt, it is men who have made the journey of self-discovery who must be relied upon to restore justice, the political counterpart of psychic balance. Thus the symbolic action, as in our American fables, has three spatial stages. It begins in a corrupt city, passes through a raw wilderness, and then, finally, leads back toward the city. But the court party is not returning to the same Milan from which it came. There is now some hope that what has been learned on the island can be applied to the world. What has been learned, needless to say, is not the lesson of primitivism. So far as the ending lends credibility to the pastoral hope, it endorses the way of Prospero, not that of Gonzalo; the model for political reform is neither Milan nor the island as they existed in the beginning; it is a symbolic middle landscape created by mediation between art and nature.

But although Shakespeare invests the pastoral hope with unexampled charm and power, in the end he refuses to make of it anything more than a hope. It is a vision that he checks, in the final scene, by unequivocal reminders of human limitations. "There is no pure comedy or tragedy in Shakespeare," says Boris Pasternak. "His style is between the two and made up of both; it is thus closer to the true face of life than either, for in life, too, horrors and delights are mixed." For all the efficacy of Prospero's magic, the ending of *The Tempest* leaves us a long way from a brave new world. There, for one thing, is Caliban, that "thing of darkness" who vows to be even wilier in the future. There, too, are those unrepentant villains, Antonio and Sebastian, who doubtless are more clever than Alonso. Prospero relinquishes power, and, what is more, he turns our thoughts to his approaching death. *Et in Arcadia Ego.* In the end we are reminded that history cannot be stopped. The pastoral design, as always, circumscribes the pastoral ideal.[34]

The pattern is remarkably like the pattern of our typical American fables. To be sure, many of them do not arrive at anything like the resolution of *The Tempest.* The American hero successfully makes his way out of society, but in the end he often is further than Prospero from envisaging an appropriate landscape of reconciliation. Nevertheless, the tacit resolution is much the same. Prospero's island community prefigures Jefferson's vision of an ideal Virginia, an imaginary land free both of European oppression and frontier savagery. The topography of *The Tempest* anticipates the moral geography of the American imagination. What is most prophetic about the play, finally, is the singular degree of plausibility that it attaches to the notion of a pastoral retreat. By making the hope so believable, Shakespeare lends singular force to its denial. *The Tempest* may be read as a prologue to American literature.

Have you still got humming birds, as in Crèvecoeur? I liked
Crèvecoeur's "Letters of an American Farmer," *so* much.
And how splendid Hermann Melville's "Moby Dick" is, &
Dana's "Two Years before the Mast." But your classic Amer-
ican literature, I find to my surprise, is *older* than our Eng-
lish. The tree did not become new, which was transplanted.
It only ran more swiftly into age, impersonal, non-human al-
most. But how good these books are! Is the *English* tree in
America almost dead? By the literature, I think it is.

D. H. Lawrence to Amy Lowell, 1916 *

It may in truth be said, that in no part of the world are
the people happier . . . or more independent than the farmers of New England.

Encyclopaedia Britannica, 1797

ALTHOUGH Shakespeare and his contemporaries
had thought about the unspoiled terrain of the New
World as a possible setting for a pastoral utopia, a fully
articulated pastoral idea of America did not emerge until
the end of the eighteenth century. The story of its emergence illustrates the turning of an essentially literary
device to ideological or (using the word in its extended
sense) political uses. By 1785, when Jefferson issued *Notes
on Virginia,* the pastoral ideal had been "removed" from
the literary mode to which it traditionally had belonged
and applied to reality.

But here again it is necessary to insist upon the vital
distinction between the pastoral ideal and the pastoral

---

* From *Amy Lowell: A Chronicle* by S. Foster Damon. Reprinted by permission
of Houghton Mifflin Company.

design. In speaking of *The Tempest* as a pastoral we refer
to a highly wrought aesthetic form, a complicated way of
ordering meanings that cannot be taken to imply any
single, clear line of action. But Jefferson's formulation of
the pastoral ideal affirms a belief which may serve as a
guide to social policy. It is impossible to exaggerate the
importance of this distinction in illuminating the obscure
borderland where so many of the confusions between art
and ideology arise. The implications of Shakespeare's play,
taken as a whole, are not at all the same as those of Prospero's "majestic vision," a kind of pastoral dream that
The Tempest encompasses. Granted that the dream seldom
has inspired more exalted expression, the play has the
effect, in the end, of checking our susceptibility to such
dreams. On the other hand, the idea that the American
continent may become the site of a new golden age could
be taken seriously in politics. As every American knows,
it has been capable of carrying an immense burden of
hope. That hope in turn has been encouraged, from the
beginning, by descriptions of the New World as a kind
of Virgilian pasture — a land depicted as if it might become the scene, at long last, of a truly successful "pursuit
of happiness."

It is necessary to stress the "as if": Elizabethan voyagers
like Captain Barlowe seldom succeeded, if indeed they ever
seriously attempted, to disguise the conventionality of the
idiom. Long after the sixteenth century, for that matter,
descriptions of America in this vein bore the unmistakable marks of their literary origins. Then, during the
eighteenth century, the situation changed. The great revolution in science and technology we associate with Sir
Isaac Newton was followed by a massive shift in prevailing
ideas about man's relations to nature. An effort was made
to rescue the pastoral — the formal literary mode — from

the confines of a decadent convention, but it failed. At the time that the old pastoral was dying, however, Europe was swept by a wave of enthusiasm for rural landscape and rural life. With this new feeling for the country came a fresh idiom, a vocabulary capable of investing the ancient ideal with new vitality.

One of the first colonial writers whose work reveals the affinity between the conditions of life in America and the pastoral ideal is Robert Beverley. His effort to interpret native experience seems to have impelled him, in spite of himself, toward a pastoral conception of society. Let us consider Beverley's *History and Present State of Virginia* (1705), the first full-scale treatment of the subject by a native.

1

Beverley's lively, if artless, book begins with the stock image of America as nature's garden, a new paradise of abundance. One gets the impression that he hit upon the motif in the same cavalier way that he became a historian. In 1703, Beverley, a self-assured, ingenuous Virginia planter, was in London on business. By chance a bookseller asked him to look over a few pages of a manuscript that dealt with Virginia, and the material struck Beverley as so inferior that he went to work at once on a book of his own. Two years later he published *The History and Present State of Virginia*. His research, such as it was, took him back to the reports of the early voyagers, and especially to an account of the new land, already mentioned, that Captain Barlowe had written for Sir Walter Raleigh. Barlowe had depicted Virginia as a natural garden of unbelievable fertility, and later Captain John Smith, drawing upon Barlowe, had elaborated upon the

same theme both in his initial description of Virginia
(1612) and in his *Generall Historie* (1624). By Beverley's
time, in any event, the convention was well established,
and he uses it to set the theme of his *History*. The
Country, says Beverley, struck the early voyagers as "so
delightful, and desirable; so pleasant, and plentiful; the
Climate, and Air, so temperate, sweet, and wholsome;
the Woods, and Soil, so charming, and fruitful; and all
other Things so agreeable, that Paradice it self seem'd
to be there, in its first Native Lustre." [1]

Although Beverley was working within an established
convention, there can be no doubt that he found it exciting. Indeed, the opening pages of the *History* might serve
as a showcase of ideas embraced by the image of America
as a new Eden. Here are no hints of uncertainty or skepticism. From the beginning Beverley identifies himself
with those worldly Elizabethan explorers who, in their
astonishment, suddenly found themselves in a pristine
garden. In expounding their views he becomes so enthusiastic that he seems to endorse the total simple logic of
Renaissance primitivism. Even the name of the colony,
he says, was selected as a tribute to the landscape (as well
as the Queen). "Virginia" refers to a land that "did still
seem to retain the Virgin Purity and Plenty of the first
Creation, and the People their Primitive Innocence. . . ."
This intoxicating idea inspires some of Beverley's most
exuberant writing. It is also his measure of value. If unimproved nature is the location of all that we desire, then
civilization as Europeans have known it can only signify
a fall or lowering of man's estate. In Beverley's words,
the Indians retained their purity because they had not
been

. . . debauch'd nor corrupted with those Pomps and
Vanities, which had depraved and inslaved the Rest of
Mankind; neither were their Hands harden'd by Labour,
nor their Minds corrupted by the Desire of hoarding up
Treasure: They were without Boundaries to their Land;
and without Property in Cattle; and seem'd to have
escaped, or rather not to have been concern'd in the first
Curse, *Of getting their Bread by the Sweat of their
Brows:* For, by their Pleasure alone, they supplied all
their Necessities; namely, by Fishing, Fowling and Hunting; Skins being their only Cloathing; and these too,
Five Sixths of the Year thrown by: Living without Labour, and only gathering the Fruits of the Earth when
ripe, or fit for use: Neither fearing present Want, nor
solicitous for the Future, but daily finding sufficient
afresh for their Subsistence.[2]

The major theme of the *History* is an elaboration of
this idea; throughout the book we hear echoes (no doubt
uncalculated) of Shakespeare's Gonzalo and, in the distance, Montaigne on cannibals:

> . . . for no kind of traffic
> Would I admit; no name of magistrate;
> Letters should not be known; riches, poverty,
> And use of service, none; contract, succession,
> Bourn, bound of land, tilth, vineyard, none . . .

In Beverley's *History,* as in *The Tempest,* the landscape
provokes a utopian vision. His ideas are grounded in a
firm doctrine of geographic determinism: "All the Countries . . . seated in or near the Latitude of *Virginia,* are
esteem'd the Fruitfullest, and Pleasantest of all Clymates.
. . . These are reckon'd the Gardens of the World. . . ."
Beverley devotes one of his four "Books" to a description
of the garden or, as he puts it, to "the Natural Product

and Conveniencies of Virginia; in its Unimprov'd State, before the English went thither." This whole section is a sustained, itemized tribute to nature's bounty. Beverley's enthusiasm occasionally leads him, as Louis Wright has noted, into the kind of inventive, high-flown boasting that was to become a hallmark of native humor. He tells, for example, of grapes so plentiful that one vine might fill a London cart, of potatoes the thickness of a child's thigh, and of a frog large enough to feed six Frenchmen. As in our Western humor, the line between fact and fancy becomes hazy. But the connection between natural facts and this hyperbolic state of mind could hardly be more plain. The whole book is colored by Beverley's exuberant, early-morning wonder in the face of nature's prodigal power:

> About Two Years ago, walking out to take the Air, I found, a little without my Pasture Fence, a Flower as big as a Tulip, and upon a Stalk resembling the Stalk of a Tulip. The Flower was of a Flesh Colour, having a Down upon one End, while the other was plain. The Form of it resembled the *Pudenda* of a Man and Woman lovingly join'd in one. Not long after I had discover'd this Rarity, and while it was still in Bloom, I drew a grave Gentleman, about an Hundred Yards, out of his way, to see this Curiosity, not telling him any thing more, than that it was a Rarity, and such, perhaps, as he had never seen, nor heard of. When we arrived at the Place, I gather'd one of them, and put it into his Hand, which he had no sooner cast his Eye upon, but he threw it away with Indignation, as being asham'd of this Waggery of Nature.[3]

So far from being ashamed of the waggeries of nature, Beverley is everywhere drawn to the lusty, spontaneous,

primitive life. He devotes one book to a detailed description of Indian culture that anthropologists still regard as
a valuable source. Although he comes out with an almost
entirely favorable impression of the natives, he is not a
doctrinaire primitivist. In the straight historical, or narrative, section of the *History,* for example, he does not shy
away from the unpleasant truth about the Indians. He
describes the massacre of the colonists in matter-of-fact
language. The Indians, he says, use cunning and surprise,
"destroying Man, Woman and Child, according to their
cruel Way of leaving none behind to bear Resentment."
And yet, in spite of such "Hellish Contrivance," when
Beverley pauses to reflect upon the origin of these bloody
episodes, he invariably puts the ultimate blame on the
aloof, superior English:

> Intermarriage had been indeed the Method proposed
> very often by the *Indians* in the beginning, urging it fre-
> quently as a certain Rule, that the *English* were not their
> Friends, if they refused it. And I can't but think it wou'd
> have been happy for that Country, had they embraced
> this Proposal: For, the Jealousie of the *Indians,* which I
> take to be the Cause of most of the Rapines and Murders
> they committed, wou'd by this Means have been alto-
> gether prevented. . . .

The quality, above all, that makes the *History* so engaging is Beverley's remarkable flexibility, his openness to
unfamiliar experience, his tolerance and even respect for
the vivacious unconstraint of the Indians. He admires their
joy in play, and he is contemptuous of the typical European sneer, especially the sanctimonious charge of sexual
license:

> The *Indian* Damsels are full of spirit, and from thence
> are always inspir'd with Mirth and good Humour. They

are extreamly given to laugh, which they do with a Grace not to be resisted. The excess of Life and Fire, which they never fail to have, makes them frolicksom, but without any real imputation to their Innocence. However, this is ground enough for the *English*, who are not very nice in distinguishing betwixt guilt, and harmless freedom, to think them Incontinent. . . .[4]

All in all, Beverley's Indians are an admirable people. They are gay, gentle, loving, generous, and faithful. And for him the reason is not far to seek. It is implicit in his controlling image, the garden landscape, and the economic fact for which it stands. The "natural Production of that Country," he says, explains the ease of life, the fabulous freedom from care, hence the charm of the natives. He sums up his reflections on primitive society by noting that natural affluence is what chiefly enables this people to get along

. . . without the Curse of Industry, their Diversion alone, and not their Labour, supplying their Necessities . . . none of the Toils of Husbandry were exercised by this happy People; except the bare planting a little Corn, and Melons, which took up only a few Days in the Summer, the rest being wholly spent in the Pursuit of their Pleasures.

This primitive utopia has an intoxicating effect upon Beverley. He believes in it. It is for him neither a hollow convention nor a wish-fulfillment fantasy. He has seen this innocent and happy people with his own eyes. And, given his assumptions about the inescapable influence of the natural environment upon the character and fortune of men, we are led to expect that the Europeans, as a result of their removal to this virgin land, quickly will be redeemed. The logic of Beverley's ruling metaphor, the new garden of the world, should result in a *History* that rises, in the end, to an inspired vision of America as paradise regained.[5]

But that is not what happens at all. On the contrary, as Beverley approaches the end he is overcome by a sense of disappointment, disgruntlement, and shame. Looking back, to be sure, one can detect a few hints of this revulsion in the earlier pages. In the narrative section, for example, he is impatient with all those Virginians and agents of the Crown who stand in the way of improvements — especially the development of towns, trade, and manufactures. And when he compares the English with the Indians, he usually takes this tone: "And indeed all that the *English* have done, since their going thither, has been only to make some of these Native Pleasures more scarce, by an inordinate and unseasonable Use of them; hardly making Improvements equivalent to that Damage." The Indians, he observes in the summary of Book III, have "reason to lament the arrival of the *Europeans,* by whose means they seem to have lost their Felicity, as well as their Innocence." Of course this critical view of the colonists could have been held within the thematic pattern of the book, especially if Beverley had shown that the British would be redeemed by the beneficent influence of the new environment. But that is not what he says. When he finally describes the influence of the new setting upon his countrymen, he becomes confused:

> In fine, if any one impartially considers all the Advantages of this Country, as Nature made it; he must allow it to be as fine a Place, as any in the Universe; but I confess I am asham'd to say any thing of its Improvements, because I must at the same time reproach my Country-Men with a Laziness that is unpardonable.[6]

In the closing pages of the *History* a paradox emerges. The new garden of the world, which Beverley has celebrated as the cause of all that is most admirable in the joyous Indian culture, now appears to have had a bad effect upon the English. The apparent contradiction, curiously enough, calls forth some of Beverley's most astonishingly vivid prose. So far from leading him to relinquish his belief in the power of nature over consciousness, his confessed ambivalence inspires him to invest the theme with even greater poetic intensity:

> . . . the extraordinary pleasantness of the Weather, and the goodness of the Fruit, lead People into many Temptations. The clearness and brightness of the Sky, add new vigour to their Spirits, and perfectly remove all Splenetick and sullen Thoughts. Here they enjoy all the benefits of a warm Sun, and by their shady Groves, are protected from its Inconvenience. Here all their Senses are entertain'd with an endless Succession of Native Pleasures. Their Eyes are ravished with the Beauties of naked Nature.

What is most remarkable about this account of the American landscape doing its work, through the senses, upon the minds of Europeans, is the degree to which Beverley anticipates the coming fashion in thought and feeling. By 1705 he apparently takes for granted the assumptions of the new sensational psychology. It is a commonplace that the emphasis of the so-called nature poets upon sensory perception and, above all, upon the influence of visible nature, had been prepared by John Locke and the theory of mind expounded in the *Essay Concerning Human Understanding* (1690). Locke was widely interpreted to mean that visual images were the primary, if not the exclusive, form in which men gained knowledge of external reality. To popularizers and literary men it seemed that Locke was identifying perception with seeing, and ideas with visual images. Whether they were in fact misinterpreting Locke's theory, as some scholars maintain, need not concern us. But cultural historians usually refer to Addison's "Pleasures of the Imagination" (1712) and James Thomson's influential poem, *The Seasons* (which began appearing in 1726), as marking the initial influence of these ideas upon aesthetic theory and poetry proper. When Beverley's *History* was published, however, Thomson was five years old, and seven years would pass before the birth of Jean-Jacques Rousseau. This is not to make any grand claim for Beverley's precocity; a new feeling for landscape, especially in painting, already was in the air. But Beverley's *History* does indicate the special appeal that the new epistemology was to have in America; it must have seemed peculiarly apt to a writer attempting to describe, say, how the British in Virginia are "ravished with the Beauties of naked Nature":

> Their Ears are Serenaded with the perpetual murmur of Brooks, and the thorow-base which the Wind plays, when it wantons through the Trees; the merry Birds too, join their pleasing Notes to this rural Consort, especially the Mock-Birds, who love Society so well, that whenever they see Mankind, they will perch upon a Twigg very near them, and sing the sweetest wild Airs in the World. . . .

Nor does nature attack only through eyes and ears:

> Their Taste is regaled with the most delicious Fruits, which without Art, they have in great Variety and Perfection. And then their smell is refreshed with an eternal fragrancy of Flowers and Sweets, with which Nature perfumes and adorns the Woods almost the whole year round.
>
> Have you pleasure in a Garden? [7]

But what is Beverley saying? The entire passage is ambiguous. Is he describing how the English have been
corrupted, led into temptation, by the incredible bounty
and charm of the new country? Or is he describing the
regenerative power of nature in Virginia? He continues
in this vein:

> Have you pleasure in a Garden? All things thrive in
> it, most surpriseingly; you can't walk by a Bed of Flowers,
> but besides the entertainment of their Beauty, your Eyes
> will be saluted with the charming colours of the Hum-
> ming Bird, which revels among the Flowers, and licks
> off the Dew and Honey from their tender Leaves, on
> which it only feeds. It's size is not half so large as an
> *English* Wren, and its colour is a glorious shining mix-
> ture of Scarlet, Green and Gold. Colonel *Byrd*, in his
> Garden, which is the finest in that Country, has a Sum-
> mer-House set round with the *Indian* Honey-Suckle,
> which all the Summer is continually full of sweet Flow-
> ers, in which these Birds delight exceedingly.

For a moment Beverley seems to offer redemption. The
image of the garden is his most evocative trope for the
joyous possibilities of life in the New World. It is not
surprising, therefore, that he comes closest to revealing
the essential cause of his ambivalence in discussing the
gardens in Virginia. "A Garden," he explains, a few pages
from the end, "is no where sooner made than there, either
for Fruits, or Flowers. . . . And yet they han't many
Gardens in the Country, fit to bear that name." [8]
If there are few gardens like Colonel Byrd's here it is
because, paradoxically, all of Virginia is a garden. The
existence of the garden-as-metaphor, nature's garden, has
hindered the appearance of gardens-in-fact. Or to put it
another way, by the closing pages of the *History* Beverley

is uncovering an ambiguity beneath the image of Virginia as a *garden*. In fact he is using the word in two distinct ways. When Beverley calls Virginia one of the "Gardens of the World," he is speaking the language of myth. Here the garden stands for the original unity, the all-sufficing beauty and abundance of the creation. Virginia is an Edenic land of primitive splendor inhabited by noble savages. The garden, in this usage, joins Beverley's own feelings with that "yearning for paradise" which makes itself felt in virtually all mythology. But when Beverley says that there are too few gardens in Virginia, he is speaking about actual, man-made, cultivated pieces of ground. This image also is an emblem of abundance, but it refers to abundance produced by work or, in Beverley's idiom, improvement. The contradiction between the two meanings of "garden" is a perfect index of the larger difference between the primitive and the pastoral ideals.[9]

Although Beverley uncovers this contradiction, he is far from comprehending it. For he is caught in no mere linguistic ambiguity.* His confusion, that is to say, arises from an inner conflict. On the one hand he is drawn to the Indians and all that they represent: a simple, effortless, spontaneous existence. They strike him as relatively autonomous, happy men. In their sensual, direct, joyous ways they call forth all the feelings that Beverley invests in the garden as a metaphor of the ideal society. But on the other hand he remains an Englishman, product of a

---

* Contemporary linguistic philosophers of the "Oxford School" do us a great service when they uncover the confusions in the words we use. But it is one thing to say that our language betrays our confused thinking, and quite another to imply that the origin of the problems we are thinking about necessarily is linguistic. In the present case, for example, the important ambiguities in Beverley's language are born of the ambiguities in his experience.

culture that values discipline, work, performance. He knows perfectly well how dangerous it is for the eyes of Europeans to be ravished by the beauties of naked nature:

> ... they depend altogether upon the Liberality of Nature, without endeavouring to improve its Gifts, by Art or Industry. They spunge upon the Blessings of a warm Sun, and a fruitful Soil, and almost grutch the Pains of gathering in the Bounties of the Earth. I should be asham'd to publish this slothful Indolence of my Countrymen, but that I hope it will rouse them out of their Lethargy, and excite them to make the most of all those happy Advantages which Nature has given them; and if it does this, I am sure they will have the Goodness to forgive me.

So ends *The History and Present State of Virginia*. It is an ingenuous, vigorous, wonderfully vivid book. The rhythm of statement and counterstatement that begins to emerge toward the end leaves us with a sense of unresolved conflict. Having begun with Nature's garden as his controlling metaphor, Beverley discovers in mid-career that he cannot accept what it implies. He does not like what has happened to the British in Virginia. He denounces them for their soft, slack ways. Yet the apparent source of this evil condition is the lush green land itself, the landscape on which his high hopes had rested at the outset. Although in the end he repudiates the convention with which he began, he is unable to define a new position. Nevertheless, we can make out the direction in which he was moving. What he wanted, after all, was to reconcile his admiration for the primitive life with what he knew of the needs of a truly civilized community. He was looking for a conception of life which would combine (to use the language of Freud) the Indians' high level of instinctual gratification with those refinements of civilization based on performance — work — hence a degree of repression. In other words, he wanted nothing less than the ideal reconciliation of nature and art which had been depicted by writers of pastoral since Virgil's time.

Beverley was groping for the distinction between two garden metaphors: a wild, primitive, or pre-lapsarian Eden in which he thought to have found the Indians, and a cultivated garden embracing values not unlike those represented by the classic Virgilian pasture. At times, in the *History*, he seems on the point of saying as much. And as a matter of fact the landscape of reconciliation, a mild, agricultural, semi-primitive terrain, was soon to become a commonplace in the rising flow of descriptive writing about America. Trying to persuade European friends and relatives to come over, or promoting various business enterprises, many colonists described the new land as a retreat, a place to retire to away from the complexity, anxiety, and oppression of European society. A favorite epithet was *asylum*, a word which also might be used to describe the setting of Virgil's first eclogue or, for that matter, Irving's Sleepy Hollow. In 1710, just five years after Beverley's *History* was published, an anonymous "Swiss Gentleman" writing from South Carolina put it this way:

> This Country, perhaps, may not abound so much with those gay and noisy amusements which generally the great and rich affect; but . . . for those who affect Solitude, Contemplation, Gardening, Groves, Woods, and the like innocent Delights of plain simple Nature . . .

this is the place! To describe the country in these terms was to resolve the root contradictions of Beverley's *His-*

tory. It is the resolution implicit in the pastoral ideal. Eighty years after the appearance of Beverley's book his contribution was acknowledged by the foremost celebrant of the pastoral theory of America.[10]

2

By 1785, when Jefferson first printed his *Notes on Virginia*, the pastoral idea of America had developed from the dim, semi-articulate compromise hinted toward the end of Beverley's *History* into something like an all-embracing ideology. But then, too, the whole tenor of Western thought had changed. During the eighty years between the two books a whole series of ideas we identify with the Enlightenment helped to create a climate conducive to Jeffersonian pastoral. I am thinking of the widespread tendency to invoke Nature as a universal norm; the continuing dialogue of the political philosophers about the condition of man in a "state of nature"; and the simultaneous upsurge of radical primitivism (as expressed, for example, in the cult of the Noble Savage) on the one hand, and the doctrines of perfectibility and progress on the other. A full discussion of the English background would require a volume in itself. Yet we can get some sense of the way the over-all shift in thought and taste contributed to the pastoral idea of America by noting three closely related preoccupations of the age: the landscape, agriculture, and the general notion of the "middle state" as the desirable, or at any rate the best attainable, human condition.

In the record of Western culture there is nothing to compare with the vogue for landscape that arose in this period. Today it is difficult to realize that Europeans have not always looked upon the landscape as an object of aesthetic interest and delight. But the fact is that landscape painting emerged as a distinct genre only during the Renaissance, and it did not achieve real popularity until the eighteenth century, when aesthetic interest in natural scenery reached something of a climax. One writer has suggested, in fact, that the arts of travel, poetry, painting, architecture, and gardening might be regarded as having been fused, in this era, into a single art of landscape. Moreover, the problem of judging the relative merits of landscapes produced a large body of aesthetic theory; complex distinctions were made between beautiful, picturesque, and sublime scenery. Nor was the landscape merely the concern of theoreticians. On both sides of the Atlantic ladies and gentlemen traveled great distances to gaze at inspiring vistas. Often they carried Claude glasses, pieces of tinted, framed glass with handles named after Claude Lorrain, who after his death in 1682 had become the most popular landscape painter of the age. When a viewer used the Claude glass the landscape was transformed into a provisional work of art, framed and suffused by a golden tone like that of the master's paintings. The significant fact is that the glass helped to create a pastoral illusion. No painter, with the possible exception of Poussin, has tried harder to depict the Virgilian ideal than Claude. His favorite subject, as described by Kenneth Clark, was a vision "of a Golden Age, of grazing flocks, unruffled waters and a calm, luminous sky, images of perfect harmony between man and nature, but touched . . . with Mozartian wistfulness, as if he knew that this perfection could last no longer than the moment in which it takes possession of our minds." Although their themes may not be as explicitly Virgilian as Claude's, a similar mood colors the work of many other popular painters of the period.[11]

Everyone knows that in the eighteenth century English poets also turned their attention as never before to the natural landscape. In the handbooks the work of men like Gay, Thomson, Akenside, Young, Gray, Goldsmith, Burns, and Cowper is often designated as "pre-romantic nature poetry." But we do not think of this school as having anything to do with the pastoral tradition, that is, with the pastoral poem defined as a distinct formal entity. Actually, reams of such poems were being turned out by English poets — in all likelihood more than in any century before or since — but they are largely unreadable: imitative, overstylized, cold, and, in a word, dead. Without question the sorry reputation of the pastoral as the dullest, most artificial of forms stems chiefly from this period. Thus Alexander Pope, who began his career in this vein:

> First in these Fields I try the Sylvan Strains,
> Nor blush to sport on *Windsor's* blissful Plains:
> Fair *Thames* flow gently from thy sacred Spring,
> While on thy Banks *Sicilian* Muses sing;
> Let Vernal Airs thro' trembling Osiers play,
> And *Albion's* Cliffs resound the Rural Lay.

Pope, it is said, was only sixteen when he wrote his pastorals, but the sad truth is that, dull as they seem, they are more accomplished and engaging than most of their kind written after 1700. Later he called them his most labored verses. What is curious, at least at first glance, is that so brilliant a technician was unable to breathe life into the pastoral at a time when the impulse toward nature manifestly was vigorous and growing in vigor.* [12]

A major difficulty lay in the very notion of provoking Sicilian Muses to song on the banks of the Thames. Even at the time poets and critics had begun to re-examine the whole conception of pastoral. By 1709, when Pope started publishing, a complicated critical debate on the subject had arisen. The neo-classicists, Pope's party, favored close imitation of the ancients; their opponents, the rationalists, wanted to modernize the mode. Starting with the reasonable assumption that the constant element in pastoral is psychological rather than formal, the rationalists (or moderns) argued for the inclusion of native materials. What they wanted was an idyllic poem after the manner of Theocritus and Virgil, but about *English* shepherds set in the *English* landscape. No doubt Spenser's work was the nearest thing to an ideal model. In the vast collection of arid theorizing on the subject, however, there is no recognition of the gulf that was opening between the convention and the feelings that people now had in the presence of nature. Accordingly the pastorals written under the banner of the moderns are on the whole indistinguishable in tone and spirit from the work of their opponents. Here, for example, are a few lines by Ambrose Philips, then generally regarded as the leading practitioner of the new style and Pope's most formidable rival:

> HOBBINOL. Full fain, O blest Eliza! would I praise
> Thy maiden rule, and Albion's golden days.
> Then gentle Sidney liv'd, the shepherds' friend:
> Eternal blessings on his shade attend!

* According to Joseph Spence, at one point Pope considered writing "American pastorals." "It might be a very pretty subject for any good genius . . . to write American pastorals; or rather pastorals adapted to several of the ruder nations, as well as the Americans. I once had a thought of writing such and talked it over with Gay. . . ."

LANQUET.   Thrice happy shepherds now: for Dorset loves
The country Muse, and our delightful groves;
While Anna reigns. Oh ever may she reign!
And bring on earth a Golden Age again.[13]

In retrospect it is not difficult to see what was happening. While the debate about the character of pastoral poetry went on and on, generating endless speculation, and while Pope and Philips and others were writing their stiff, bookish poems, the old form actually was dying. At the very same time enthusiasm for the landscape was rising on all sides and Addison was laying a theoretical groundwork for a new visual literature. In his 1712 *Spectator* papers on the "Pleasures of the Imagination" he expounds what is in effect an aesthetic corollary to Locke's system. Arguing that sight is the most perfect of the senses and the true source of the pleasure we get from the exercise of the imagination, Addison prepares the way for the modern (romantic) notion of the imagination as a way of knowing. On the spectrum of modes of perception, with the gross sensual pleasures such as touch at one pole, and the refinements of the pure intellect (understanding) at the other, he locates the imagination in the middle position. It mediates between the brutish senses and the abstract intelligence. Although he grants to the powers of the understanding the highest place, he also notes certain ways in which the imagination is more effective and, by implication, more important: it offers obvious, easily acquired pleasures. "It is but opening the eye," he says, "and the scene enters." [14]

What kind of scene? What objects are likely to provide the greatest pleasure? If Addison had to choose, he would choose works of nature over works of art. No mere art, he says, can provide that sense of "vastness and immensity"

that we get from nature itself. Yet he does not identify
the natural with the wild. In fact, he says, "we find the
works of nature still more pleasant, the more they resemble those of art." It is the mingling of mind with brute
matter that is most affecting. "Hence it is that we take delight in a prospect which is well laid out, and diversified
with fields and meadows; woods and rivers; . . . in any
thing that hath such a variety or regularity as may seem the
effect of design, in what we call the works of chance." Ap-
plying this principle to taste in gardens, Addison rejects
the formal English style, with its geometrically sculptured
trees, in favor of the more "natural" style prevailing in
France and Italy. In those countries, he says, "we see a
large extent of ground covered over with an agreeable
mixture of garden and forest, which represent every where
an artificial rudeness, much more charming than that neat-
ness and elegancy which we meet with in . . . our own
country." [15]

In effect Addison is building a theoretical bridge between the ideal of the old pastoral, the imaginary landscape of reconciliation, and a new attitude toward the
environment more congenial to a scientific and commercial age. Indeed, the "mixture of garden and forest"
with its "artificial rudeness" points toward the pastoral
image of America. Addison extends the notion of the
garden to larger and larger tracts of land until he seems
to be talking about the whole rural scene as if it were one
vast garden. The formal style of garden which he rejects
embodies a purely aristocratic, leisure-class ideal of conspicuous waste. It separates beauty from utility and work.
"It might, indeed, be of ill consequence to the public," he notes, "as well as unprofitable to private persons, to alienate so much ground from pasturage, and

the plow. . . ." Instead, he proposes a garden landscape
that is in fact a farmland:

> But why may not a whole estate be thrown into a kind
> of garden by frequent plantations, that may turn as much
> to the profit, as the pleasure of the owner? A marsh over-
> grown with willows, or a mountain shaded with oaks, are
> not only more beautiful, but more beneficial, than when
> they lie bare and unadorned. Fields of corn make a pleas-
> ant prospect, and if the walks were a little taken care of
> that lie between them, if the natural embroidery of the
> meadows were helpt and improved by some small addi-
> tions of art, and the several rows of hedges set off by trees
> and flowers, that the soil was capable of receiving, a man
> might make a pretty landscape of his own possessions.[16]

Within a few years James Thomson was following Addison's lead. Although his work cannot be described as
pastoral in the old, strict meaning of the term, there is a
more important sense in which it is continuous with the
Virgilian tradition. To Pope, or, for that matter, to
Philips, pastoral was a name for a fixed body of poetic
conventions, and they had presided over its death. But if
we take the vital element in pastoral to be the design, the
ordering of meaning and value around the contrast between two styles of life, one identified with a rural and
the other with an urban setting, then the pastoral was
by no means dead. On the contrary, Thomson was helping to save the mode by fashioning a new idiom, a language closer to the actual feelings that men had about
nature. Although strongly influenced by Virgil, Thomson
had the wit to dispense with many of the formal devices
of traditional pastoral and address himself to experience.

At once, array'd
In all the colours of the flushing year

> By Nature's swift and secret-working hand,
> The garden glows, and fills the liberal air
> With lavish fragrance; while the promis'd fruit
> Lies yet a little embryo, unperceiv'd,
> Within its crimson folds. Now from the town,
> Buried in smoke, and sleep, and noisome damps,
> Oft let me wander o'er the dewy fields,
> Where freshness breathes, and dash the trembling drops
> From the bent bush, as through the verdant maze
> Of sweetbriar hedges I pursue my walk;
> Or taste the smell of dairy . . .

To our taste the language of "The Seasons" (1730) may seem excessively literary, but we can hardly fail to recognize a new resonance, an awareness of topography actually perceived that rarely, if ever, gets into the stock pastorals of the age. It seems "something strange," said Joseph Warton in 1756, speaking of Pope, "that in the pastorals of a young poet there should not be found a single rural image that is new." The crucial difference lies in the particularity of Thomson's imagery: the town buried in smoke, the walk through a maze of sweetbriar hedges, the smell of cattle strong enough to taste. Gone are the "flowery meads," "verdant vales," "mossy banks," and "shady woods" of the standard poetic landscape of the age. Although we recognize a stock pastoral contrast here, most of Thomson's contemporaries would have denied that "The Seasons" was a pastoral poem. Actually, the poem epitomizes an early stage in the revitalizing of the pastoral design that would culminate in Wordsworth's *The Prelude*. In these lines Thomson prefigures a characteristic Wordsworthian topography: a town of "noisome damps" set against a country landscape. What adds to the significance the passage holds for us is the fact that young

Thomas Jefferson copied most of it into his commonplace book.[17]

In recent years scholars have clarified the relation between the vogue for landscape as an aesthetic object and the great scientific revolution that began in the seventeenth century. In the first place, as knowledge of the physical universe rapidly increased, a new sort of environmentalism was encouraged in every department of thought and expression. The psychology of Locke demonstrated how meaning was derived, through sense perception, from the surrounding world of objects. Yet the scene "out there" no longer was conceived as it had been in Shakespeare's time. Telescopes and microscopes were uncovering a vast, orderly cosmos behind the visible mask of nature. It is hardly surprising, therefore, that a new mode of feeling, a heightened responsiveness to the environment now developed. In a sense the change in aesthetic attitudes toward nature was as revolutionary as the change in science. For now a well-composed landscape, whether depicted in words or paint, might arouse some of the feelings that men had when they contemplated the grand design of the Newtonian universe.

Although scientific knowledge seemed to drain certain traditional religious myths of their cogency and power, so that it no longer was quite possible to read Genesis as it once had been read, the same knowledge enabled artists to invest the natural world with fresh mythopoeic value. The movements of the heavenly bodies, space (an awesome, unimaginable infinity of space), and the landscape itself all were to become repositories of emotions formerly reserved for a majestic God. It was not enough to call this newly discovered world beautiful; it was sublime.

But the conventions of the old pastoral provided a totally inadequate vehicle for such ideas and emotions. The pastoral, said Dr. Johnson, is an "easy, vulgar, . . . disgusting" mode; whatever images it supplied were "long ago exhausted." Meanwhile, a large audience was being instructed in the appreciation of the landscape as a great religious metaphor, an expositor (in Emerson's fine phrase) of the divine mind.[18]

In the same period the importance of landscape was enhanced by a new veneration for certain of its sociological or economic meanings. There was nothing new, to be sure, about the expression of agrarian pieties. In England, much earlier, the general renaissance of classical learning had helped to intensify native respect for agriculture. Any number of Greek and Latin authors besides Virgil provided models of praise for husbandry: Hesiod, Xenophon, Cato, Cicero, Varro, and Horace among them. Then, too, the glamorous prospect of settling new lands, remote from centers of civilization, added to the appeal of rural as against urban life. The stock literary contrast between the happiness and innocence of a bucolic golden age and the corrupt, self-seeking, and disorderly life of the city (or court) had been a ruling motif of Elizabethan literature. Often it was invoked with high seriousness, as an apparent affirmation of a pastoral ethic that Sidney had compressed into a tidy couplet:

> Greater was that shepheards treasure,
> Then this false, fine, Courtly pleasure.

Although we generally assume that this simple moral scheme is offered ironically, many readers then and now have taken it straight. According to Hallett Smith, for example, most Elizabethan writers of pastoral were seriously committed to the whole system of value which turns on the "rejection of the aspiring mind." Whether they were or not, the fact remains that they provided a paradigm for the agrarian celebration. By substituting the husband-man for the shepherd in Smith's account of the pastoral ethic, it was easy to transform the farmer into a cult fig-ure. Instead of striving for wealth, status, and power, he may be said to live a good life in a rural retreat; he rests content with a few simple possessions, enjoys freedom from envying others, feels little or no anxiety about his property, and, above all, he does what he likes to do.[19]

To this traditional literary motif the eighteenth cen-tury added a new set of theoretical arguments. Political economists and agricultural reformers now dwelled as never before upon the primacy of agriculture in creating the wealth of nations. While the physiocrats, extremists of the movement, insisted that husbandry was the *only* true source of economic value, most of the experts, including the incomparable Adam Smith, agreed that agriculture was the primary and indispensable foundation of national prosperity. All of these ideas contributed to the steadily rising prestige of farmers and farming, often reaching the public by way of agricultural reformers and popularizers such as Arthur Young.

But there also was a curious strain of extravagance run-ning through the cult, a seemingly neurotic tendency that these rational theories cannot explain. After the middle of the century, among the upper classes, the taste for the bucolic rose to an extraordinary pitch of faddish excite-ment. A passion for gardening and playing farmer cropped up in remote villages of England as well as at the court of Louis XVI. Some of the fever seems to have been pro-

voked by a recognition, however inchoate, of the impending threat to the supremacy of rural values. The early
encroachment of what we would call industrialization becomes obvious, for instance, in Oliver Goldsmith's popular poem, "The Deserted Village." Published about the
time the craze reached its peak, in 1770, the poem is a
lament for the charms of rural life which the Enclosure
Acts apparently were destroying. The whole situation is
remarkably like that of Virgil's first eclogue. Here again
the idyllic mood is intensified by the hostile policies of
the state; like Virgil, Goldsmith dwells upon the eviction
of the "laboring swain" from "lovely bowers of innocence
and ease." Part of the effect is achieved by glossing over
the distinction between the countryman who actually does
the work and the gentleman (or poet) who enjoys rural
ease. But the point, after all, is to idealize a rural way of
life. By having the swain, like a traditional literary
shepherd, enjoy both the ease of the rich and the simple
honesty of the poor, Goldsmith is able to suggest (as William Empson has put it) a "beautiful relation" between
the two classes. What is important about the rural world,
in any event, is not merely the agricultural economy but
its alleged moral, aesthetic, and, in a sense, metaphysical
superiority to the urban, commercial forces that threaten
it. Both the tone and the thematic design of "The Deserted
Village" will reappear in American responses to industrialization. As a matter of fact, Goldsmith foresees the
migration of the rural muse:

> Ev'n now, methinks, as pond'ring here I stand,
> I see the rural virtues leave the land.
> Down where yon anch'ring vessel spreads the sail . . .

The vessel, needless to say, is bound for America, and
when it sails "sweet Poetry" (that "charming nymph")

will be aboard. In the New World Englishmen will have
another chance to realize the village ideal, a social order
represented neither by the hideous wilderness:

> Those pois'nous fields with rank luxuriance crown'd,
> Where the dark scorpion gathers death around;
> Where at each step the stranger fears to wake
> The rattling terrors of the vengeful snake;

nor by the European city where the swain flees only

> To see profusion that he must not share;
> To see ten thousand baneful arts combin'd
> To pamper luxury, and thin mankind . . .[20]

A belief in the superiority of rural life was a sociological
corollary, for Goldsmith's generation, to the widely accepted ethical doctrine that the "middle state" was the
best of all possible human conditions. According to this
venerable idea, man is the creature who occupies the middle link in the "great chain of being," the point of transition between the lower and higher, animal and intellectual
forms of being. A grand, compelling metaphor, it figures
a moral scheme which Arthur Lovejoy calls the "ethic of
the middle link." Men, in this view, must accept an unsatisfactory but nonetheless unavoidable compromise between their animal nature and their rational ideals.
Whether we like it or not, the theory goes, we will always
find ourselves mediating between these contraries, and so
we had best learn to live in the uncomfortable middle.*

---

* It is worth noting that the unavoidable relativism implicit in this
"middle state ethic" probably contributes to the pragmatic bent of American thought in the nineteenth century. By definition this is a scheme
that admits of no absolute solutions, and looks to an endless series of
ad hoc decisions, compromises, and adjustments in resolving problems.

It is a moral position perfectly represented by the image of a rural order, neither wild nor urban, as the setting of man's best hope.[21]

Although it stems from classical antiquity, the "middle state" theory enjoyed its greatest diffusion and acceptance during the eighteenth century. One thing that made it so popular was the need to reconcile increasingly strong claims by extremists at either pole of the debate about the nature of man. The two schools may be called progressivist and primitivist. At the same time that electrifying advances in learning and the arts seemed to sanction a belief, or at least a new degree of confidence, in the perfectibility of civilized man, books like Beverley's *History* were supplying Europe with apparent testimony in favor of primitive life. In effect, they called into question the whole value of civilization. After all, the Virginia Indian did seem happier and better-natured than the average Londoner. In many ways the stock figure of the Noble Savage, which now became immensely popular, resembles the good shepherd of the old pastorals. At least they share certain negative virtues: neither displays the undignified restlessness, ambition, or distrust so common in advanced societies; neither, above all, is consumed by a yearning for approval and praise.[22]

Yet, as we have seen in the case of Beverley, it was not easy for intelligent men to maintain a primitivist position. In his *History* Beverley anticipates (in a crude and half-articulate fashion, to be sure) the dialectic that was to be worked out again and again during the next few generations. On a higher plane of sophistication, Jean-Jacques Rousseau was drawn to the spontaneity and freedom he associated with primitive life; but he too had to face the undeniable fact that "natural man" was, by European standards, amoral, uncreative, and mindless. Unable finally to endorse either the savage or the civilized model, Rousseau was compelled to endorse the view that mankind must depart from the state of nature — but not too far.* He came to believe, as Lovejoy says, that " 'perfectibility' up to a certain point was desirable, though beyond that point an evil." [23]

The desirability of a similar reconciliation between the animal and rational, natural and civilized, conditions of man always had been implied by the pastoral ideal. In his influential lectures on rhetoric (published in 1783), Hugh Blair, the Scottish divine, made the connection between pastoral and the "middle state" theory explicit. Discussing the nature of pastoral, Blair follows the line taken by Pope in his "Discourse on Pastoral Poetry" (written in 1704), distinguishing between the two ways of depicting bucolic life. One might be called realistic: shepherd life described as it actually is, which is to say, "mean, servile, and laborious." The second way is to impute the polished taste and cultivated manners of high civilization to the swain. But if the first is repellent, the ideas of real shepherds being what they are ("gross and low"), the second is implausibly idealistic; hence Blair rejects them both.

> Either of these extremes is a rock upon which the Poet will split, if he approach too near it. We shall be disgusted if he give us too much of the servile employments

---

* Curiously enough, Rousseau thought that mankind had passed through the ideal state during the pastoral phase of cultural evolution, by which he meant a pastoral situation in a literal, anthropological sense: a society of herdsmen.

and low ideas of actual peasants . . . and if . . . he makes his Shepherds discourse as if they were courtiers and scholars, he then retains the name only, but wants the spirit of Pastoral Poetry.

He must, therefore, keep in the middle station between these. He must form to himself the idea of a rural state, such as in certain periods of Society may have actually taken place, where there was ease, equality, and innocence; where Shepherds were gay and agreeable, without being learned or refined; and plain and artless, without being gross and wretched. The great charm of Pastoral Poetry arises from the view which it exhibits of the tranquillity and happiness of a rural life. This pleasing illusion, therefore, the Poet must carefully maintain.

Although Blair is talking about the best setting for a pastoral poem, not about social actualities, he does leave the way open for readers who might be tempted to confuse the two. At certain times, he says, such ideal societies of the "middle landscape" actually may have existed. And if they existed once, why not again? The conjecture is worth mentioning because of the great prestige that Blair's work was to enjoy in America. "I shall take care to get Blair's lectures for you as soon as published," wrote Thomas Jefferson to James Madison in 1784. The *Lectures* were used as a textbook in American colleges down to the middle of the next century.[24]

Attractive as it was, the idea of a society of the middle landscape was becoming less easy to believe in during the 1780's. In England the process of "improvement," or what we should call economic development, already seemed to have gone too far. By then the enclosures were destroying the vestiges of the old, rural culture, and the countryside was cluttered with semi-industrial cities and dark, satanic mills. At this juncture the next thought was obvious and irresistible. For three centuries Englishmen had been in the habit of projecting their dreams upon the unspoiled terrain of the New World. Long before the 1780's George Herbert had got the prevailing attitude into a couplet:

> Religion stands on tip-toe in our land,
> Ready to passe to the *American* strand.

Andrew Marvell, who probably was indebted to Captain John Smith's account in *The Generall Historie of Virginia*, put it this way in his "Bermudas":

> What should we do but sing his Praise
> That led us through the watry Maze,
> Unto an Isle so long unknown,
> And yet far kinder than our own?
> . . . .
> He lands us on a grassy Stage;
> Safe from the Storms, and Prelat's rage.
> He gave us this eternal Spring,
> Which here enamells every thing;
> And sends the Fowl's to us in care,
> On daily Visits through the Air.
> He hangs in shades the Orange bright,
> Like golden Lamps in a green Night.

In America it still was not too late (or so one might imagine) to establish a home for rural virtue. In the period of the Revolution, accordingly, the pastoral idea of America caught on everywhere in England. To take one of many examples, here is Richard Price, the outspoken Unitarian minister, friend of Franklin and Jefferson, and a leading British champion of the American cause:

The happiest state of man is the middle state between the *savage* and the *refined*, or between the wild and the luxurious state. Such is the state of society in CONNECTICUT, and some others of the *American* provinces; where the inhabitants consist, if I am rightly informed, of an independent and hardy YEOMANRY, all nearly on a level — trained to arms . . . clothed in homespun — of simple manners — strangers to luxury — drawing plenty from the ground — and that plenty, gathered easily by the hand of industry; and giving rise to early marriages, a numerous progeny, length of days, and a rapid increase — the rich and the poor, the haughty grandee and the creeping sycophant, equally unknown — protected by laws, which (being their own will) cannot oppress . . . O distinguished people! May you continue long thus happy; and may the happiness you enjoy spread over the face of the whole earth. . . .

During the winter of 1785, when he received his copy of Price's *Observations on the . . . American Revolution* containing this apostrophe, Jefferson was getting the manuscript of *Notes on Virginia* ready for a Paris printer. "I have read it with . . . pleasure," he wrote to Price, "as have done many others to whom I have communicated it. The spirit which it breathes is as affectionate as the observations . . . are wise and just." Needless to say, there is no question here of Jefferson's having been influenced by Price's version of the pastoral ideal; by this time it was in the air on both sides of the Atlantic.[25]

In 1785, for example, a pamphlet entitled *The Golden Age* was published. Although it does not reveal either the author or the place of publication, the title page itself is a nice exhibit of the affinities between America and the pastoral ideal:

THE
GOLDEN AGE:
OR,
Future Glory of North-America
Discovered
By An Angel to Celadon,
in
Several Entertaining Visions.

VISION I.
———— *Ferrea*
Definet, ac toto surget gens Mundo
Virgil Eclog IV.
Thus Englished,
The iron past, a golden Age shall rise,
And make the whole World happy, free, and wise.

B Y   C E L A D O N

Printed in the Year, M, DCC, LXXXV.

The sixteen-page pamphlet begins with a description of
Celadon, a "man strictly honest, and a real lover of the
country," who lives in "one of the American States." He
had fought gloriously against England during the war and
now he is "anxious to know the future condition of his
people." One day, while musing on this subject, he is at
last blessed with an agreeable vision. At this point, the
writer shifts to the first person, Celadon describing his experience in his own words. [26]

On a summer evening Celadon is tired, and he walks
out and sits on the bank of a stream where he is "delighted
with the music of the groves." With murmuring waters
below and sighing winds above he falls into a trance. An

angel appears and all "nature seemed to smile at his approach." Describing himself as America's "guardian," the angel announces that he has come to dissolve Celadon's doubts about the future. The rest of the pamphlet consists of a dialogue between Celadon and his celestial confidant, who provides a graphic description of the American millennium. At the climax the angel takes Celadon in his arms and carries him to a high mountain in the center of the land. From the great summit the entire continent is visible. To the east he sees "spacious cities . . . thriving towns . . . a thick conjunction of farms, plantations, gardens . . . laden with every kind of fruit!"; it is a countryside "charmingly diversified . . . plentifully watered" and filled with "elegant buildings adapted to all the purposes of life." Then the angel orders Celadon to turn his face westward, and he is "equally surprised at the wide extended landscape. This western part of America," he says, "is as yet but an uncultivated desart; the haunt of savages; and range of wild beasts. — But the soil in general is much richer than that of the eastern division. . . ." Into this great land the "poor, the oppressed, and the persecuted will fly . . . as doves to their windows." The pamphlet ends with a vision of "a beauteous world rising out of a dreary wilderness," unlimited progress (England's decline having begun with the American Revolution), the taming of the Indians, the conversion of the Jews, and the American way spreading over the face of the earth. Although more extravagant than most versions, *The Golden Age* is in essence another statement of the pastoral idea of America.

Two years earlier the same idea had provided the controlling theme of a full-scale, on-the-spot interpretation of American experience. In his popular *Letters from an*

American Farmer, J. Hector St. John de Crèvecœur (another of Jefferson's correspondents) weaves all of the thematic strands just discussed into a delightful, evocative, though finally simple-minded, book. It projects the old pastoral ideal, now translated into a wholly new vocabulary, on to the American scene. At the outset Crèvecœur sets up a typical pastoral situation. Writing as a "humble American Planter, a simple cultivator of the earth," he deferentially addresses a learned Englishman who has invited the rustic American to become his correspondent. Here the writer is no mere observer or dreamer; he *is* the American husbandman — a fact which underscores the literalness of the pastoral ideal in a New World setting. "Who would have thought," the first letter begins, "that because I received you with hospitality and kindness, you should imagine me capable of writing with propriety and perspicuity?" Incredulity is the dominant note of the opening pages. At first, says the epistolary narrator, he was overwhelmed by the invitation: is it possible, he had asked, that a man who has lived in that "big house called Cambridge" where "worldly learning is so abundant," actually wants to receive letters from a plain farmer? The problem had troubled him. He had consulted his wife and his minister. She laughed at him, said the "great European man" obviously was not in earnest; on the other hand, the minister reassured him, arguing that he need only write in his everyday, spoken language. If the letters be not "elegant," he said, at least they "will smell of the woods, and be a little wild." Besides, might the American rustic not "improve" the sophisticated Englishman by acquainting him with the "causes which render so many people happy?" At last the farmer was persuaded; but even then, he exclaims, "on recollecting the difference

between your sphere of life and mine, a new fit of astonishment seized us all!" [27]

Anyone who knows American writing, incidentally, will
be reminded of all the later fictional narrators who begin
in the same way, impulsively dissociating themselves from
the world of sophistication, Europe, ideas, learning, in a
word, *the world,* and speaking in accents of rural ignorance. "I was young and ignorant," says Mark Twain's
narrator at the beginning of *Roughing It,* compressing
the gist of innumerable American literary beginnings.
But Crèvecœur, alas, makes no use of the potentially complex framework he establishes at the outset. Although he
might have provided a counterpoise to the farmer's innocence, the Englishman virtually drops out of sight, and
Crèvecœur loses control of the initial contrast between the
farmer and his learned friend. The *Letters* exhibit one of
the great hazards awaiting American writers who attempt
to work in this convention. To put it much too simply
for the moment, the trouble is that under the singularly
beneficent circumstances of native experience, the barrier
between art and reality is likely to break down. Crèvecœur,
as one critic notes, "lived a kind of pastoral poetry." Having adopted the point of view of the self-derogating, uneducated American swain, he seems to forget that it is a
literary device. (Actually, his family belonged to the lesser
nobility of France, and he had been educated in a Jesuit
school.) After the sophisticated Englishman drops out of
sight the farmer's simple definition of reality is allowed
to govern the entire work.[28]

To account for the peculiarly "modern, peaceful and
benign" qualities of his rural life, the farmer introduces
all of the familiar environmentalist assumptions of the
age. He believes that men everywhere are like plants, de-

riving their "flavor" from the soil in which they grow. In America, with its paucity of established institutions, however, the relation between mankind and the physical environment is more than usually decisive. Geography pushes men into farming, which is of course the noblest vocation. But the land is significant not only for the material and political benefits it confers; at bottom it determines everything about the new kind of man being formed in the New World. "What should we American farmers be," he asks, "without the distinct possession of that soil? It feeds, it clothes us, from it we draw even a great exuberancy. . . ." [29]

The delightfully graphic notion of man drawing exuberancy from the soil exhibits Crèvecœur's imagination at its best; back of it lies his absolute confidence in the power of imagery. Skeptical of abstract ideas, he is a kind of homespun Lockian who thinks of the land, or rather the landscape, as an object that penetrates the mind, filling it with irresistible pictures of human possibilities. To paraphrase Addison, he has but to open his eyes and the American ethos enters. Just to see this virgin terrain is to absorb the rudiments of a new consciousness, the American "philosophy," as he calls it. Without the sense of the landscape as a cardinal metaphor of value, the *Letters* could not have been written. Indeed, for the farmer it is the metaphoric even more than the physical properties of land which regenerate tired Europeans by filling them to overflowing with exuberancy. We are reminded of Robert Beverley's exuberant style, not to mention Melville's and Whitman's — Whitman, whose hero will move from the contemplation of a single spear of grass to his barbaric yawp. It is not surprising that Crèvecœur was one of the writers who convinced D. H. Lawrence that only the

"spirit of place" really can account for the singular voice
we hear in American books. In the *Letters,* as elsewhere
in our literature, the voice we hear is that of a man who
has discovered the possibility of changing his life. Landscape means regeneration to the farmer. In sociological
terms, it means the chance for a simple man, who does
actual work, to labor on his own property in his own behalf. It gives him a hope for the leisure and economic
sufficiency formerly — which is to say, in Europe — reserved for another class. It therefore represents the possibility of a secular, egalitarian, naturalistic "resurrection"
(as Crèvecœur calls it at one point), the religious content
of which is now deeply embedded in a new pastoral idiom.
"These images I must confess," says the farmer of rural
scenes, "I always behold with pleasure, and extend them
as far as my imagination can reach: for this is what may
be called the true and the only philosophy of an American farmer." [30]

The farmer's imagination, as it turns out, has a long
reach. With his pleasant farm at the center, he expands
the topographical image endlessly, until it achieves mythic
magnitude. Eastward it reaches to Europe, encompassing
l'ancien régime, an oppressive social order of "great lords
who possess everything, and of a herd of people who
have nothing." Westward he extends it to the dark forest
frontier where something "very singular" happens to
Europeans. Their lives being "regulated by the wildness
of the neighbourhood," they become "ferocious, gloomy
and unsociable." As he describes the frontiersmen, they
are "no better than carnivorous animals of a superior
rank, living on the flesh of wild animals" — native American Calibans. Like the Europeans rebuked by Beverley,
they show what happens to those who trust too much in

the "natural fecundity of the earth, and therefore do little." Like lazy Virginians, they do not work, do not
improve the natural environment. But now, when the
typical immigrant arrives in America, he no longer beholds the raw wilderness. On the contrary, he sees "fair
cities, substantial villages, extensive fields, an immense
country filled with decent houses, good roads, orchards,
meadows, and bridges, where an hundred years ago all was
wild, woody, and uncultivated!" This ideal landscape obviously has little in common with the "plantation"
dream of Shakespeare's Gonzalo or the Edenic Virginia
of Beverley's *History*. The point needs to be made because Crèvecœur, like many another American writer,
often is mistaken (in Lawrence's derisive phrase) for a
"littérateur-Child-of-Nature-sweet-and-pure." But he is
nothing of the sort. He does not believe, as Lawrence
says he does, that Nature is sweet and pure. He admires
improved nature, a landscape that is a made thing, a
fusion of work and spontaneous process. "This formerly
rude soil," he explains, "has been converted by my father
into a pleasant farm, and in return it has established all
our rights." [31]

Taken as a whole, the moral geography of the *Letters*
forms a neat spatial pattern, a compelling triptych that
figures an implied judgment upon all the conditions of
man which may be thought to exist between the savagery
of the frontier on one side and the court of Versailles on
the other. It is in this sense a potentially mythic idea, an
all-encompassing vision that converts the ethic of the middle link into "the true and the only" philosophy for
Americans. No one possibly could miss the main point;
nevertheless, the farmer dramatizes it in the final letter,
called "Distresses of a Frontier Man." By now the rev-

olutionary war has begun, and he writes in a mood of great agitation. Exposed to imminent Indian attack, desperately anxious for the safety of his family, torn between loyalty to the homeland and to his neighbors, depressed about the need to abandon his farm with all its improvements, he reluctantly has decided to move further west to a peaceful Indian village. But the decision is a painful one. Perhaps, he says, "I may never revisit those fields which I have cleared, those trees which I have planted, those meadows which, in my youth, were a hideous wilderness, now converted by my industry into rich pastures and pleasant lawns." After weighing the relative advantages of moving east and moving west, he chooses west. Like many exponents of the pastoral theory of America, he veers toward the primitive. But even as he makes his choice, he reaffirms the ideal of the middle landscape:

> I will revert into a state approaching nearer to that of nature, unencumbered either with voluminous laws, or contradictory codes, often galling the very necks of those whom they protect; and at the same time sufficiently remote from the brutality of unconnected savage nature. Do you, my friend, perceive the path I have found out? it is that which leads to . . . the . . . village of ———, where, far removed from the accursed neighbourhood of Europeans, its inhabitants live with more ease, decency, and peace, than you imagine: where, though governed by no laws, yet find, in uncontaminated simple manners all that laws can afford. Their system is sufficiently complete to answer all the primary wants of man, and to constitute him a social being, such as he ought to be in the great forest of nature.

In the *Letters from an American Farmer* there is no mention of Arcadia, no good shepherd, no stock of poeti-

cisms derived from Virgil, no trite antithesis of country
and town, no abstract discord between Nature and Art.
And yet all of these traditional features of the pastoral
mode are present in new forms supplied by American experience. Instead of Arcadia, we have the wild yet potentially bucolic terrain of the North American continent;
instead of the shepherd, the independent, democratic husbandman with his plausible "rural scheme"; instead of the
language of a decadent pastoral poetry, the exuberant
idiom, verging toward the colloquial, of the farmer; and
instead of generalized allusions to the contrast between
country and town, Crèvecœur begins to explore the difference between American and European cultures, a complex variation of the grand Nature-Art antithesis which
informs all pastoral literature. As for the dream of reconciliation, it is now reinforced by the agrarian philosophy
and the "middle state" theory of the age, and it thereby
takes on a credibility it never before had possessed.

Today, looking back across the great gulf created by industrialism, we can easily see what was wrong with the
pastoral theory of America. We say that it embodied a
naïve and ultimately static view of history, and so it did.
But to project this judgment into the past is to miss the
compelling power of the ideal in its eighteenth-century
context. That is why we so often mistake it for a primitivist fantasy. From our perspective they may look equally
regressive, but the distinction between them was once a
vital element in the American consciousness, and to ignore
it is to blur our sense of the past. Crèvecœur was no primitivist; he had no illusions about the condition of man in a
state of nature, nor did he repudiate change or economic
development. But how, then, we may well ask, did he
reconcile his admiration for roads, bridges, fair cities and

improvements of all sorts with the bucolic ideal? What did he think would happen when the new society approached that delicate point of equilibrium beyond which further change, which is to say, further departure from "nature," would be dangerous?

Part of the answer lies in the way he conceived of America's immediate situation. Here, as he saw it, was a fringe of settlements on the edge of an immense, undeveloped, and largely unexplored continent. At the time nine out of ten Americans were farmers living in a virtually classless society, and all of the best informed statesmen and political economists agreed that agriculture would remain the dominant enterprise of the young nation for centuries to come. To be sure, in retrospect we can see that industrialization already had begun in England, but no one at the time conceived of the process even remotely as we do. (It would be roughly fifty years before the word "industrialism" came into use.) Thus when Crèvecœur's farmer makes one of the unavoidable inventories of feudal institutions missing in the new world — "no aristocratical families, no courts, no kings, no bishops" and so on — he includes "no great manufacturers employing thousands." This sentence could only have been written on the other side of the gulf created by industrialization. Obviously Crèvecœur is unaware of any necessary relation between large-scale manufacturing and changing social institutions; the connections between technology, economic development, and the thrust of deprived people for a higher living standard, which we take for granted, did not exist for him; as he thinks about America's future, it involves nothing of that irreversible and accelerating process of change now regarded as the very powerhouse of history. As a result, he is able to imagine a society which will embrace both the pastoral ideal and the full application of the arts, of power. His farmer is enlisted in a campaign to dominate the environment by every possible means. "Sometimes," he says, "I delight in inventing and executing machines, which simplify my wife's labour." If there is a time ahead when the natural bent of Americans for improvement will have to be curbed, it is too remote to think about. "Here," says the farmer, ". . . human industry has acquired a boundless field to exert itself in — a field which will not be fully cultivated in many ages!" [32]

But the unimproved landscape is only part of the answer. Granted that native conditions lent a remarkable degree of credibility to his idea of America's future, there is a more important reason for Crèvecœur's failure to recognize the obvious dilemma of pastoral politics. Indeed, his desire to avoid such problems is precisely what made the rural scheme so attractive in the first place. Throughout the *Letters* the farmer describes America as the "great asylum," a "refuge" from Europe, from power struggles, politics, or, in our sense of the word, from history itself. In spite of all the deceptive local details, it is still possible to identify the main features of the idyllic Virgilian landscape in this idealized picture of America. It is a place apart, secluded from the world — a peaceful, lovely, classless, bountiful pasture.

3

With the appearance of Crèvecœur's *Letters* in 1782 the assimilation of the ancient European fantasy to conditions in the New World was virtually complete. None of the obvious devices of the old pastoral could be detected in the farmer's plausible argument for an American rural scheme. And yet, essentially the same impulse was at work beneath the surface; it manifested itself in the static, anti-historical quality of the whole conception, and by 1785 Jefferson said as much. Even as he endorsed the same goals he recognized the problems Crèvecœur had glossed over. How could a rural America possibly hold off the forces already transforming the economy of Europe? What policies would a government have to adopt to preserve a simple society of the middle landscape? Even if they were adopted, what chance of success would they have when, minor distinctions aside, the same sort of men who were coming to power in England already were busy in America?

To appreciate the quality of Jefferson's answers it is first necessary to be clear about his own preferences. If he had been a mere practical politician, a man never tempted by utopian impulses, his criticism of the goals endorsed by Crèvecœur would not be of much interest. But Jefferson was an ardent exponent of much the same theory; he argued for it with a poetic intensity rare in the writings of American statesmen. A man of immeasurably larger mind than Crèvecœur, Jefferson also located the weak spot — the place where the theory and the facts were likely to come apart. This did not cause him to repudiate either; rather he kept his hold upon both, using the facts of power to check a proclivity toward wishful thinking, and avoiding the shallows of simple pragmatism by an insistence upon the need for theory — for long-range ideals. In fact, the "doubleness" of Jefferson's approach toward the pastoral ideal is akin to the basic design of the literary mode. Not being an artist, however, he never had to get all of his feelings down in a single place. As a result, we have to piece together his "version of pas-

toral," a way of ordering values which carries us far beyond Crèvecœur's unqualified affirmation — beyond it in
the direction of the complex attitudes at work in the
writings of, say, Henry Thoreau and Robert Frost.

Still, Jefferson's initial commitment is unambiguous.
Nowhere in our literature is there a more appealing, vivid,
or thorough statement of the case for the pastoral ideal
than in *Notes on Virginia*. Although it is his only original,
full-length work, it has never been widely read. One need
only turn the pages to see the reason: at first glance it
looks like a cross between a geography textbook and a
statistical abstract. Written in answer to a series of questions from François Marbois, a French diplomat, the
book has a dense, dry, fact-laden surface. This should not
be allowed to obscure its enduring qualities: it is in its way
a minor classic, rising effortlessly from topographical fact
to social analysis and utopian speculation. In 1780, when
Jefferson began writing, the Revolution was not yet over
and the ultimate goals of the new Republic were very
much on his mind. Hence, he not only describes the way
things are in Virginia, but he allows himself to reflect
upon the way they might be.

Here again in *Notes on Virginia* the land — geography
— has a decisive function. Nothing could be more remote
from the never-never land of the old pastoral than the
setting Jefferson establishes at the outset. This is the opening sentence of the book:

> Virginia is bounded on the East by the Atlantic: on
> the North by a line of latitude, crossing the Eastern
> Shore through Watkins Point, being about 37°. 57'.
> North latitude; from thence . . .

Nor is this matter-of-fact tone confined to the first few sentences or even pages. Of the twenty-three "queries" into

which the book is divided, the first seven are devoted to
the natural environment:

Query	I	Boundaries of Virginia
Query	II	Rivers
Query	III	Sea-Ports
Query	IV	Mountains
Query	V	Cascades
Query	VI	Productions Mineral, Vegetable and Animal
Query	VII	Climate

In choosing to begin the *Notes* in this fashion, Jefferson
almost certainly had no complicated literary theory in
mind. Marbois had sent him twenty-two questions. He
simply rearranged them in a logical order, expanding
their number to twenty-three, and now he was answering
them one by one. Nevertheless, the sequence embodies
the "Lockian" presuppositions, already discussed, about
the relations between nature, man, and society. Because
they are part of the book's structure, moreover, these ideas
do some of their most effective literary work indirectly.
Thus the wealth and specificity of the geographical detail
in the first part of the book conditions our response to
what comes later, much as the whaling lore in *Moby-Dick*
affects our feelings about the metaphysical quest. Before
we get to Query XIX, with its passionate defense of a
rural society, the image of the rich, rugged, but largely
undeveloped, terrain of Virginia has been firmly embedded
in our minds. It helps to make credible, as no abstract
argument could, Jefferson's feeling for the singular plas-
ticity of the American situation.

The book contains repeated movements from fact to
feeling. Thus the fissure beneath the Natural Bridge is
first described in mathematical terms, "45 feet wide at

the bottom, and 90 feet at the top. . . . Its breadth
in the middle, is about 60 feet, but more at the ends
. . ." and so on. And then, with hardly a break: "It is
impossible for the emotions, arising from the sublime, to
be felt beyond what they are here: so beautiful an arch,
so elevated, so light, and springing, as it were, up to heaven, the rapture of the Spectator is really indiscribable!"
The treatment of landscape in the *Notes* recalls a phrase
from John Locke's second treatise on government, a book
which influenced Jefferson as much as any other single
work: "in the beginning," said Locke, "all the world was
America. . . ." A great hope makes itself felt almost
wordlessly in the texture of *Notes on Virginia*. The
topographical details, like the opening scenes of *The
Tempest,* establish a firm naturalistic base for utopian
revery.[33]

Like Shakespeare's Gonzalo, moreover, Jefferson at
times seems to be carried away by the promise of a fresh
start. Were it made a question, he says at one point,
speaking of the rarity of crime among the Indians,

> . . . were it made a question, whether no law, as
> among the savage Americans, or too much law, as among
> the civilized Europeans, submits man to the greatest evil,
> one who has seen both conditions of existence would
> pronounce it to be the last: and that the sheep are
> happier of themselves, than under the care of the wolves.

It is true that this statement, taken out of context, does
sound as if Jefferson had joined a simple-minded cult of
Nature. But as so often in American writing, what first
strikes us as the merest vagary of a primitivist proves, on
closer inspection, to be something quite different. (Admittedly, there is a charming absurdity in the very prospect
of the Virginia *philosophe* casting his lot with the In-

dians.) Jefferson is not endorsing the Indian style of life here. What appears as a preference for the primitive actually is a rhetorical device: *were* it made a question, he says, he would prefer the Indian way, the whole point being that in the New World such questions need not arise.[34]

Jefferson's syntax, in other words, is aimed at a pastoral, not a primitivist, affirmation. (Depicting the people as sheep is a stock device of pastoral poetry.) We might call it the "syntax of the middle landscape," a conditional statement which has the effect of stressing a range of social possibilities unavailable to Europeans. Jefferson remained fond of this sentence structure all his life. As late as 1812, for instance, he is still telling John Adams (speaking of the depravity of France and England) that "if science produces no better fruits than tyranny, murder, rapine and destitution of national morality, I would rather wish our country to be ignorant, honest and estimable as our neighboring savages are." Adams, incidentally, missed or pretended to have missed the point; in his reply he refers, as if Jefferson had made an unconditional statement, to his friend's "Preference to Savage over civilized life." But that is not what Jefferson meant. True, this syntax leads the mind toward an affirmation of primitive values, cutting it loose from an easy acceptance of established ideas and institutions, but behind it there is no serious intention of going all the way. It is a perfect expression of the American pastoral ethos. Actually, Jefferson had few illusions about the felicities of primitive life, as he makes clear elsewhere in *Notes on Virginia:*

> The women are submitted to unjust drudgery. This I believe is the case with every barbarous people. With such, force is law. The stronger sex . . . imposes on the

weaker. It is civilization alone which replaces women in the enjoyment of their natural equality. That first teaches us to subdue the selfish passions, and to respect those rights in others which we value in ourselves.

The symbolic setting favored by Jefferson, needless to say, resembles the Virgilian landscape of reconciliation, but it now is a real place located somewhere between *l'ancien régime* and the western tribes.[35]

Moreover, it is a landscape with figures, or at least one figure: the independent, rational, democratic husbandman. The eloquent passage in Query XIX in which he makes his appearance is the *locus classicus* of the pastoral ideal as applied to America. For a century afterwards, politicians, orators, journalists, and spokesmen for the agricultural interest cherished Jefferson's words and, above all, the ringing declaration: "Those who labour in the earth are the chosen people of God, if ever he had a chosen people, whose breasts he has made his peculiar deposit for substantial and genuine virtue." Sentences lifted from Query XIX are among the most frequently quoted but least understood of Jefferson's words. Taken out of context, for one thing, the picture of the husbandman loses a good deal of the charm and meaning it owes to the force of its opposite. It is significant that Jefferson creates this ideal figure not while discussing agriculture, but rather when he is considering the place of manufactures in the new nation. Although the process of industrialization, in our sense of the word (the use of machine power in large-scale manufacturing) had not begun, the states had stepped up the productivity of domestic manufactures during the war. When it was over there were some who believed that the foundation for a permanent system of manufactures had been laid. This is the prospect that alarms Jefferson and that sets off the well-known tribute to rural virtue as the moral center of a democratic society.

At the outset, it is true, Jefferson asserts that his countrymen "will certainly return as soon as they can" to their former dependence upon European manufactures. But it is unlikely that he was so confident as he pretends. By 1785, as we shall see, he was making quite another prediction. Besides, the vehemence of this attack upon the idea of native manufactures would seem, in itself, to belie his words. A close look at the structure and style of his answer to Query XIX (*The present state of manufactures, commerce, interior and exterior trade?*) is revealing. This short chapter falls neatly into two parts, each written in a distinct style, each exemplifying one of Jefferson's attitudes toward America's future. It is important enough and, fortunately, brief enough to cite in its entirety. Here is the first part:

> We never had an interior trade of any importance. Our exterior commerce has suffered very much from the beginning of the present contest. During this time we have manufactured within our families the most necessary articles of cloathing. Those of cotton will bear some comparison with the same kinds of manufacture in Europe; but those of wool, flax and hemp are very coarse, unsightly, and unpleasant: and such is our attachment to agriculture, and such our preference for foreign manufactures, that be it wise or unwise, our people will certainly return as soon as they can, to the raising raw materials, and exchanging them for finer manufactures than they are able to execute themselves.
>
> The political œconomists of Europe have established it as a principle that every state should endeavour to

manufacture for itself: and this principle, like many others, we transfer to America, without calculating the difference of circumstance which should often produce a difference of result. In Europe the lands are either cultivated, or locked up against the cultivator. Manufacture must therefore be resorted to of necessity not of choice, to support the surplus of their people. But we have an immensity of land courting the industry of the husbandman. *Is it best then that all our citizens should be employed in its improvement, or that one half should be called off from that to exercise manufactures and handicraft arts for the other?* [Italics added.]

So far Jefferson is speaking in the voice of a scientific rationalist. In arguing that America is an exception to the general rule that nations should aim at economic self-sufficiency, he merely endorses the commonplace views of the political economists — notably Adam Smith. The argument has the incidental appeal, for Jefferson, of underscoring the geographic basis of American politics: abundance of land is what makes America a special case. But is it best for America to remain an exclusively rural nation? It is just at this point that he betrays the "soft center" of the pastoral theory. Having worked his way up to the great question in the spare, dispassionate idiom of political economy, he suddenly changes voices. Between two sentences, a rhetorical question and an "answer," he abandons discursive reasoning. Here is the question, once again, and the rest of Query XIX:

Is it best then that all our citizens should be employed in its improvement, or that one half should be called off from that to exercise manufactures and handicraft arts for the other? Those who labour in the earth are the chosen people of God, if ever he had a chosen people, whose breasts he has made his peculiar deposit for substantial and genuine virtue. It is the focus in which he keeps alive that sacred fire, which otherwise might escape from the face of the earth. Corruption of morals in the mass of cultivators is a phænomenon of which no age nor nation has furnished an example. It is the mark set on those, who not looking up to heaven, to their own soil and industry, as does the husbandman, for their subsistence, depend for it on the casualties and caprice of customers. Dependence begets subservience and venality, suffocates the germ of virtue, and prepares fit tools for the designs of ambition. This, the natural progress and consequence of the arts, has sometimes perhaps been retarded by accidental circumstances: but, generally speaking, the proportion which the aggregate of the other classes of citizens bears in any state to that of its husbandmen, is the proportion of its unsound to its healthy parts, and is a good-enough barometer whereby to measure its degree of corruption. While we have land to labour then, let us never wish to see our citizens occupied at a workbench, or twirling a distaff. Carpenters, masons, smiths, are wanting in husbandry: but, for the general operations of manufacture, let our work-shops remain in Europe. It is better to carry provisions and materials to workmen there, than bring them to the provisions and materials, and with them their manners and principles. The loss by the transportation of commodities across the Atlantic will be made up in happiness and permanence of government. The mobs of great cities add just so much to the support of pure government, as sores do to the strength of the human body. It is the manners and spirit of a people which preserve a republic in vigour. A degeneracy in these is a canker which soon eats to the heart of its laws and constitution.

Although the term *agrarian* ordinarily is used to describe the social ideal that Jefferson is endorsing here, to call it *pastoral* would be more accurate and illuminating.

This is not a quibble: a serious distinction is involved. To begin with, the astonishing shift from the spare language of political economics to a highly figurative, mythopoeic language indicates that Jefferson is adopting a literary point of view; or, to be more exact, he is adopting a point of view for which an accepted literary convention is available. Notice the conventional character of the second half of this passage. There can be no doubt about the influence of literary pastoral upon Jefferson. We have touched on his youthful reading of James Thomson, and his love of Virgil and the classical poets is well known. In 1813 he still was quoting Theocritus in Greek to John Adams. But quite apart from his direct response to pastoral poetry, we are more likely to get a precise idea of Jefferson's meaning here if we interpret it as the expression of a pastoral rather than an agrarian ideal.[36]

What, then, is the difference? The chief difference is the relative importance of economic factors implied by each term. To call Jefferson an agrarian is to imply that his argument rests, at bottom, upon a commitment to an agricultural economy. But in Query XIX he manifestly is repudiating the importance of economic criteria in evaluating the relative merits of various forms of society. Although the true agrarians of his day, the physiocrats, had demonstrated the superior efficiency of large-scale agriculture, Jefferson continues to advocate the small, family-sized farm. Ordinarily, he does not think about farms as productive units, a fact which may help to account for his own dismal financial record as a farmer. He is devoted to agriculture largely as a means of preserving rural manners, that is, "rural virtue." Therefore, he would continue shipping raw materials to European factories (and back across the Atlantic in finished form) no

matter what the monetary cost. Unlike the fully committed agrarians, he admits that an agricultural economy
may be economically disadvantageous. But that does not
trouble him, because he rejects productivity and, for that
matter, material living standards, as tests of a good society.
The loss of what nowadays would be called "national in-
come," he explains, "will be made up in happiness and
permanence of government."

All of this makes more sense once we recognize the noble
husbandman's true identity: he is the good shepherd of
the old pastoral dressed in American homespun. One of
that traditional figure's greatest charms always had been
his lack of the usual economic appetites. As Renato Poggioli observed, in the literary mode the shepherd appears
everywhere as the "opposite of the *homo oeconomicus.*"
To find an almost perfect model of the Jeffersonian economy we need only recall the situation of the happy rustic in
Virgil's first eclogue. There too the economy makes possible the contained self-sufficiency of the pastoral community. Like the swain in his bucolic retreat, the Virginia
farmer on his family-sized farm would produce everything that his family needs and at most a little more. The
goal is sufficiency, not economic growth — a virtual stasis
that is a counterpart of the desired psychic balance or
peace. (Notice that here "permanence of government"
is one of Jefferson's chief concerns.) By equating desires
with needs, turning his back on industry and trade, the
husbandman would be free of the tyranny of the market.
Here the absence of economic complexities makes credible
the absence of their usual concomitant, a class structure. Jefferson grounds this happy classless state in the
farmer's actual possession of land; in such a society all
men would adopt an aloof patrician attitude toward ac-

quisitive behavior. (At this point the Lockian right of private property is woven into the pattern of these "natural" principles.) But the physical attributes of the land are less important than its metaphoric powers. What finally matters most is its function *as a landscape* — an image in the mind that represents aesthetic, moral, political, and even religious values. Standing on his farm, "looking up to heaven" and down to his "own soil and industry," Jefferson's rustic is in a position to act out all of the imperatives of the "middle link" ethic. Moreover, he derives these principles in much the same way that Crèvecœur's farmer drew exuberancy from the landscape. Jefferson invokes the familiar botanical metaphor when he describes the "germ of virtue" being "deposited" in the husbandman's breast. It is this moral seed which enables him to renounce nascent industrial capitalism, that is, a market-regulated society in which men must submit to the "casualties and caprice of customers." [37]

In developing the contrast between the American farmer and the European workman, Jefferson also follows the Virgilian pattern. He sets the joy of independence against the misery of dependence. As in the first eclogue, the crux here is the relation that each man enjoys with the natural environment — the land. Like Tityrus, the husbandman's life is attuned to the rhythms of nature; but the European worker's access to the land, like Melibœus', has been blocked. As a result he becomes economically dependent, a condition which "begets subservience and venality" and so "suffocates the germ of virtue." That is what will happen, Jefferson is warning, if America develops its own system of manufactures. The fresh, health-giving, sunlit atmosphere of Virginia will be replaced by the dark, foul air of European cities. At a time when the idea of the market economy as the "natural" regulator of economic behavior enjoys great prestige, Jefferson associates it with oppressive institutions. So far from being part of benign nature, it is a kind of malformation or disease. It generates "mobs" of bestialized men who will eat, like a "canker," at the heart of the republic. At this point the intensity of Jefferson's feeling, a welling up of physical revulsion, makes itself felt in the imagery of suffocation, degeneracy, and disease ("sores on the body") with which he depicts this awful prospect. What concerns him is not agriculture for its own sake, but rather small-scale yeoman farming as the economic basis for what may be called a desirable general culture. It is the "happiness," "manners and spirit" of the people — the over-all quality of life — that rules out manufactures. These concerns have more in common with Virgil's poetry than with *The Wealth of Nations*.

But the literary ancestry of the husbandman, like that of Crèvecœur's farmer, is obscured by the peculiar credibility imparted to the pastoral hope under American conditions. In this environment the political implications of the convention are subtly changed. William Empson has shown that the old pastoral, "which was felt to imply a beautiful relation between rich and poor," served as a mask of political purposes. By the "trick" of having a simple shepherd express strong feelings in a sophisticated language, the poet managed to combine the best of both worlds. The reader in turn was put in a position to enjoy the vicarious resolution of one of his culture's deepest conflicts.* The result was a kind of conservative quietism. In a sense, Jefferson's idealized portrait of the husband-

---

* I am paraphrasing Empson's argument, but I feel that Empson fails to clarify the distinction between those versions of pastoral which enable the reader to enjoy an easy resolution of the conflict and those which enforce the poet's ironic distance from the pastoral dream, that is, between sentimental and complex pastoralism.

man, a simple yet educated man, a noble democrat, per-
forms the same service. But with this vital difference: Jefferson is supposed to be talking about an actual social
type in an actual society; at no point is the reader warned,
as he inevitably is when he picks up a poem, that he is
crossing the boundary between life and art. Jefferson's
enemies, who liked to denounce him as a dreamer,
might have underscored their abuse by noting the resemblance between his husbandman and the literary shepherd. Later this "mythical cult-figure" (Empson's term)
of the old pastoral will reappear as the Jacksonian "common man." If this democratic Everyman strikes us as a
credible figure, it is partly because the pastoral ideal has
been so well assimilated into an American ideology. He
achieves the political results outside literature formerly
achieved by the shepherd. In the age of Jackson there no
longer will be any need to insinuate a beautiful relation
between the rich and the poor. By his mere presence
the "common man" threatens — or promises — to supplant them both.[38]

In the egalitarian social climate of America the pastoral
ideal, instead of being contained by the literary design,
spills over into thinking about real life. Jefferson extends
the root contrast between simplicity and sophistication to
opinions on every imaginable subject. "State a moral case,"
he says, "to a ploughman and a professor. The former will
decide it as well, and often better than the latter, because
he has not been led astray by artificial rules." The pastoral
element in this famous anti-intellectual homily is apparent. The true American is the ploughman, whose values
are derived from his relations to the land, not from "artificial rules." To be sure, it may be said that the shepherd's
moral superiority always had been based upon a some-

what obscure metaphysical link with "nature." But the conviction with which Jefferson makes this kind of asser- tion stems from his belief in the unspoiled American landscape as peculiarly conducive to the nurture of the "moral sense." It disseminates germs of virtue.[39]

All of these feelings are at work in Jefferson's warning against sending an American youth to Europe, far from the benign influences of his native ground:

> He acquires a fondness for European luxury and dissipa- tion and a contempt for the simplicity of his own coun- try; he is fascinated with the privileges of the European aristocrats, and sees with abhorrence the lovely equality which the poor enjoy with the rich in his own country: he contracts a partiality for aristocracy or monarchy; he forms foreign friendships which will never be useful to him, and loses the season of life for forming in his own country those friendships which of all others are the most faithful and permanent: he is led by the strongest of all the human passions into a spirit for female in- trigue destructive of his own and others happiness, or a passion for whores, destructive of his health, and in both cases learns to consider fidelity to the marriage bed as an ungentlemanly practice and inconsistent with hap- piness: he recollects the voluptuary dress and arts of the European women, and pities and despises the chaste af- fections and simplicity of those of his own country. . . .

It would be difficult to find a more comprehensive state- ment of the pastoral ethic in American terms. Away from his own country the chain that links the young man's manners to his vocation and his vocation to the soil is broken. A little further on, Jefferson extends the argu- ment to literary style. An American educated abroad, he warns, is likely to return

> . . . speaking and writing his native tongue as a for-
> eigner, and therefore unqualified to obtain those distinc-
> tions which eloquence of the pen and tongue ensures
> in a free country; for I would observe to you that what
> is called style in writing or speaking is formed very early
> in life while the imagination is warm, and impressions
> are permanent. I am of opinion that there never was
> an instance of man's writing or speaking his native
> tongue with elegance who passed from 15. to 20. years
> of age out of the country where it was spoken.

Implying that the peculiar "simplicity" of American manners will be embodied in a native language and therefore
a distinct style, Jefferson points toward the work of Henry
Thoreau, Mark Twain, Ernest Hemingway, and Robert
Frost — all of whom would write "versions of pastoral"
in a distinctively American idiom.[40]

In Jefferson's time this way of thinking about American
differences reflected the popular mood of post-Revolution-
ary patriotism. It was fashionable, during the 1780's, to
take a pious tone toward Europe, sophistication, aristocracy, luxury, elegant language, etc. "Buy American!" was
a slogan that arose from the premises of pastoral economic
theory. It meant that crude local products were preferable
(on moral grounds, of course) to European finery. Years
ago Constance Rourke named these attitudes, which appear everywhere in native lore, the fable of the contrast.
She borrowed the name from Royall Tyler's play, *The
Contrast*, first produced in 1787. Setting a forthright, plain-
spoken American (named Manly) against a foolish, foppish,
and sinister Anglicized type (named Dimple), Tyler fashioned a trite, melodramatic action. But the character who
really took hold of the popular imagination, and who
ultimately was to reappear in the guise of Uncle Sam,

was a Yankee rustic named Jonathan. His great charm is an intriguing blend of astuteness and rural simplicity. Beyond question he is a comic cousin of Jefferson's noble husbandman. Combining country manners with a wit capable of undoing city types, he too embodies the values of the middle landscape. As a matter of fact, Constance Rourke's seminal book, *American Humor* (1931), is a study of the pastoral impulse in its most nearly pure indigenous form. It suggests the degree to which the cultural situation in Jefferson's America, with its rare juxtaposition of crudity and polish, may have approximated the environment from which the pastoral convention had arisen centuries before. Perhaps that explains D. H. Lawrence's impression — see the epigraph to this chapter — that American writers somehow had by-passed English literature, had reached behind it for their inspiration, thereby producing work that seems "older" than English writing.[41]

Almost immediately after the first printing of *Notes on Virginia,* however, Jefferson referred to the idea of a permanently undeveloped, rural America as "theory only, and a theory which the servants of America are not at liberty to follow." He made the statement in answering some questions from G. K. van Hogendorp, who had been reading the *Notes.* Possibly Jefferson's skepticism was encouraged by Hogendorp's shrewd characterization of Dr. Price in the same letter as an "enthousiast" who was "little informed of local circumstances" and who was "forming a judgement on America rather from what . . . [he wished] mankind to be, than from what it is." In any event, Hogendorp asked Jefferson to elaborate upon his ideas about the new nation's future economic development.

Here is Jefferson's reply:

> You ask what I think on the expediency of encouraging
> our states to be commercial? Were I to indulge my own
> theory, I should wish them to practice neither commerce
> nor navigation, but to stand with respect to Europe
> precisely on the footing of China. We should thus avoid
> wars, and all our citizens would be husbandmen.

He admitted that under certain conditions America might
some day be forced to engage in commerce and manufactures.

> But that day would, I think be distant, and we should
> long keep our workmen in Europe, while Europe should
> be drawing rough materials and even subsistence from
> America. But this is theory only, and a theory which the
> servants of America are not at liberty to follow. Our
> people have a decided taste for navigation and com-
> merce. They take this from their mother country: and
> their servants are in duty bound to calculate all their
> measures on this datum . . .[42]

Right here the reason for the strange break in the middle of Jefferson's reply to Query XIX becomes apparent.
In this letter he reverts to the political economist's language, tone, and point of view. Once again the scientific
empiricist is speaking. He tells Hogendorp that much as
he would like to calculate measures according to his own
theory, there is a datum which the theory cannot encompass. What could be further from the high-flown rhetoric
of the apostrophe to "those who labor in the earth"? It
is as if Jefferson had taken that leap into rhetoric because
he knew where the discursive style led. To put the pastoral
theory of America into effect it would be necessary at
some point, in fact almost immediately, to legislate against
the creation of a native system of manufactures. But to

curb economic development in turn would require precisely the kind of governmental power Jefferson detested. Moreover, he admits that the American people already have a "decided taste" for business enterprise. The seeds of a dynamic, individualistic economic system had been imported from Europe. Hence the noble husbandman inhabits a land of "theory only," an imaginary land where desire rules fact and government is dedicated to the pursuit of happiness.

What are we to make of this apparent contradiction? Ever since Jefferson's death scholars have been trying to discover order in — or to impose it upon — his elusive, unsystematic thought, but without much success. It simply does not lend itself to ordinary standards of consistency. There is no way, finally, to cancel out either his ardent devotion to the rural ideal or the cool, analytic, pragmatic tone in which he dismisses it. All of his commitments embrace similar polarities. He praises the noble husbandman's renunciation of worldly concerns, but he himself wins and holds the highest political office in the land; he is drawn to a simple life in a remote place, but he cherishes the fruits of high civilization — architecture, music, literature, fine wines, and the rest; he wants to preserve a provincial, rural society, but he is devoted to the advance of science, technology, and the arts. How, then, can we evaluate Jefferson's conflicting statements about the pastoral ideal?

The first step in understanding Jefferson, as Richard Hofstadter suggests, is to dispense with shallow notions of consistency. Observing that he "wanted with all his heart to hold to the values of agrarian society, and yet . . . believed in progress," Hofstadter concludes that such deep ambiguities lie at the very center of his temperament. To charge him with inconsistency, after all, is to imply that a firm grasp of the facts or the rigorous imposition of logic might have improved the quality of his thought. But that is a mistake. The "inconsistencies" just mentioned are not the sort that can be swept aside by a tidying-up of his reasoning. They are not mere opinions. They stem from a profound ambivalence — a complex response to the conflicting demands of the self and society.[43]

It is striking to see how closely Jefferson's attitude toward his own role conforms to the pattern of contradiction displayed in Query XIX. What he most wanted, or so he always maintained, was to live quietly in his rural retreat at Monticello. As Roland Van Zandt has shown, Jefferson thought of the whole realm of power politics, war, and conflict as something Americans might have ignored but for the evil pressures of the great (European) world. Invariably he regards his political activity as a temporary departure from the natural and proper pattern of his life. It is always a concession to an emergency.[44]

At the beginning of the Revolution, in 1775, he writes to John Randolph to explain how much he hates "this unnatural contest." Even at the age of thirty-two he is contemplating a permanent retirement from active life.

> My first wish is a restoration of our just rights; my second a return of the happy period when, consistently with duty, I may withdraw myself totally from the public stage and pass the rest of my days in domestic ease and tranquillity, banishing every desire of afterwards even hearing what passes in the world.

In 1793, after serving as governor of Virginia, minister to France, and Secretary of State, he still regards political life as a diversion.

> When I first entered on the stage of public life (now twenty-four years ago), I came to a resolution never to

> engage while in public office in any kind of enterprise
> for the improvement of my fortune, nor to wear any
> other character than that of a farmer.

And, again, in 1795:

> It is now more than a year that I have withdrawn myself
> from public affairs, which I never liked in my life, but
> was drawn into by emergencies which threatened our
> country with slavery, but ended in establishing it free.
> I have returned, with infinite appetite, to the enjoyment
> of my farm, my family and my books, and . . . deter-
> mined to meddle in nothing beyond their limits.

A year later he was a candidate for the Presidency; twelve
years of public service ensued, and it was not until 1809
that Jefferson finally retired to his rural seat at Monticello.
Calls to public duty followed by withdrawals to the country comprise the very rhythm of his life. Beyond question
Jefferson's contradictory attitudes toward the pastoral ideal
are analogous to these conflicting private motives.[45]

When he expresses a yearning for country peace and
pleasure he sometimes veers toward sentimental pastoralism. Then he insists that his true vocation is farming;
that a rural life is the proper and the "natural" life for
an American, and that he only emerges from his sylvan
retreat when a nefarious "other" threatens the peace. Seen
from this vantage the forces which make politics necessary
are not truly American; they always originate somewhere
else. Projected upon the larger scene, this attitude implies
that all America is — or should be — a kind of pastoral
oasis. There is the implication that the young Republic,
if it only could remain undisturbed, would not have to
confront any serious problems of power. This way of con-
ceiving reality is buttressed by Lockian environmentalism,
a theory of knowledge and behavior which on the whole

neglects individual or psychic sources of aggression. It is
the feudal past, perpetuated by corrupt, repressive institutions, that accounts for the evil economic and military
depradations of Europe. If the pristine landscape of Virginia is conducive to the nurture of democratic values,
think of what happens to men in the European terrain,
with its dark, crowded cities, its gothic ruins, and crowded
workshops!

A cluster of natural images — the rural landscape,
plants, organic growth, gardens, and gardening — connects
Jefferson's political theories with his physical tastes. Like
a great poet, he often achieves a striking fusion of intellectual and sensual experience. "I have often thought,"
he writes to Charles Wilson Peale in 1811, "that if heaven
had given me choice of my position and calling, it should
have been on a rich spot of earth, well watered, and near
a good market for the productions of the garden. No occupation is so delightful to me as the culture of the earth,
and no culture comparable to that of the garden." A
garden is a miniature middle landscape. It is a place as
attractive for what it excludes as for what it contains.
Working in his garden he is free of the anxieties and
responsibilities of political office. By the same token, if
all America somehow could be transformed into a garden,
a permanently rural republic, then its citizens might escape
from the terrible sequence of power struggles, wars, and
cruel repressions suffered by Europe. This is the "logic"
back of what is known as the Jeffersonian dream — a
native version of an ancient hope.[46]

But, again, the impulse to withdraw from the world
represents only one side of Jefferson. The other side, represented by his active political role, his empirical estimate
of social forces, emerges clearly in his gradual accommo-

dation to the idea of American manufactures. In a brilliant letter to Benjamin Austin, in 1816, he retraces his thinking on the subject. "You tell me I am quoted," he writes, "by those who wish to continue our dependence on England for manufactures. There was a time when I might have been so quoted with more candor, but within the thirty years which have since elapsed, how are circumstances changed!" He then recounts the history of the Napoleonic era, the wars, the exclusion of American shipping from the seas, and, all in all, the experience of what, in 1785, "we did not then believe, that there exist both profligacy and power enough to exclude us from the field of interchange with other nations: that to be independent for the comforts of life we must fabricate them ourselves. We must now place the manufacturer by the side of the agriculturist." But what about the question he had raised in Query XIX? Did not the immensity of unimproved land in America argue the more or less permanent supremacy of agriculture?

> The former question is suppressed, or rather assumes a new form. Shall we make our own comforts, or go without them, at the will of a foreign nation? He, therefore, who is now against domestic manufacture, must be for reducing us either to dependence on that foreign nation, or to be clothed in skins, and to live like wild beasts in dens and caverns. I am not one of these . . .[47]

The controlling principle of Jefferson's politics is not to be found in any fixed image of society. Rather it is dialectical. It lies in his recognition of the constant need to redefine the "middle landscape" ideal, pushing it ahead, so to speak, into an unknown future to adjust it to ever-changing circumstances. (The ideal, in fact, is an abstract embodiment of the concept of mediation between the extremes of primitivism and what may be called "over-civilization.") In 1816, accordingly, the momentous question he had raised in 1785 — in our language: shall America industrialize? — necessarily "assumes a new form." Why? Because circumstances have changed. A quarter of a century earlier it had been possible to imagine, at least when the wishful side of his temperament was in charge, that the choice was still open: America might still have elected "to stand, with respect to Europe, precisely on the footing of China." But this is no longer feasible. In 1816 the choice for America is continuing economic development or one of two unacceptable alternatives: becoming a satellite of Europe or regressing to the life of cavemen.

To accept the need for manufactures in 1816, therefore, does not mean abandoning his basic principles. But it does mean that he must accommodate himself, however reluctantly, to the factory system he detests. It is important to stress his reluctance: one cannot read the long sequence of Jefferson's letters on the subject without recognizing the painful anxiety that this accommodation arouses in him. But then it may be said that his acceptance of the need for manufactures hardly constitutes mediation between ideal and reality — mediation, that is, in the literal sense of establishing a compromise; with this decision, after all, the distance between the two seems to be opening rather than closing. So it is. That is why the feasible policy in 1816 includes what had seemed avoidable in 1785. To admit this is only to underline the irony with which Henry Adams portrays the career of Jefferson: during his eight-year term as President of the United States, Jefferson's policies had the effect of creating precisely the kind of society he did not want. But then, even in 1785, Jefferson had acknowledged in his wistful letter to Hogendorp that America was not headed in the direction he preferred. By the time he put it in words, Jefferson already had begun to shift his vision of a pastoral utopia from the future to the past.[48]

In his greatest moments, Jefferson was clear about the difference between his preferences and what he called "circumstances." Similarly, he defined many of the important decisions of his life as choices between private pleasures and public responsibilities — between the garden and the world. So thoroughgoing was the doubleness of his outlook that he virtually had a separate language for each side. Far from being a handicap, it seems, this inconsistency was a source of his political strength. Tracing the history of the Jeffersonian image since 1826, Merrill Peterson attributes his continuing fascination for us largely to the fact that he expresses decisive contradictions in our culture and in ourselves.[49]

Beginning in Jefferson's time, the cardinal image of American aspirations was a rural landscape, a well-ordered green garden magnified to continental size. Although it probably shows a farmhouse or a neat white village, the scene usually is dominated by natural objects: in the foreground a pasture, a twisting brook with cattle grazing nearby, then a clump of elms on a rise in the middle distance and beyond that, way off on the western horizon, a line of dark hills. This is the countryside of the old Republic, a chaste, uncomplicated land of rural virtue. In his remarkable book, *Virgin Land*, Henry Nash Smith shows that down to the twentieth century the imagination of Americans was dominated by the idea of transforming the wild heartland into such a new "Garden of the World." On the whole, Americans were unsentimental about unmodified nature. Attempts to idealize the wilderness, moreover, finally produced only a sterile, formula-ridden art — the "western story." As Tocqueville noted, the wilderness was precious to most Americans chiefly for what could be made of it — a terrain of rural peace and happiness.

This symbolic landscape did not exist only on canvas or in books, or, for that matter, in the minds of those who were familiar with art and literature. What makes *Virgin Land* so illuminating is that Smith demonstrates in detail how the goals embodied in this image governed behavior on many planes of consciousness. It activated the stubborn settlers who struggled for years to raise crops in what was literally a desert; it led congressmen to insist upon certain impractical provisions of the Homestead Act; it lay back of the peculiarly bitter frustration of western farmers beginning in the 1870's; it kept alive the memory of Thomas Jefferson; it caused artists and writers both popular and serious to lose touch, as time went on, with social realities; it excited the imagination of Frederick Jackson Turner, not to mention all the historians who so eagerly endorsed the "frontier hypothesis" as the most plausible explanation for the Americanness of various attitudes and institutions — and one could go on. "The image of this vast and constantly growing agricultural society," says Smith, ". . . became . . . a collective representation, a poetic idea . . . that defined the promise of American life. The master symbol of the garden embraced a cluster of metaphors expressing fecundity, growth, increase, and blissful labor in the earth. . . ." [50]

For more than a century, then, the American people held on to a version of the pastoral ideal not unlike the one that Jefferson had set forth in 1785, investing it with a quality of thought and feeling that can only be called mythic. The "master symbol" described by Smith is not a static emblem of private felicity; it is the aim of a grand collective enterprise. Accordingly, when he calls it the *myth* of the garden, Smith is not using the term as it is used either in ordinary discourse, meaning an illusory idea, or in contemporary literary criticism, as applied for example to the poetry of Yeats, meaning a manifestly fictitious, sophisticated invention designed to place a cosmological frame around one poet's vision of life. As he uses the term, myth is a mode of belief. He is saying that Americans, so far as they shared an idea of what they were doing as a people, actually saw themselves creating a society in the image of a garden. The quality of their feeling was not unlike that of the Greeks in the Homeric poems, for whom myth presumably expressed a believable definition of reality.

Yet Jefferson, as we have seen, could not give full credence to the myth. Although he never entirely repudiated it, he knew perfectly well that it did not encompass all of the essential truth about American experience. In detached moods, he recognized the restless striving of his countrymen, their get-ahead, get-rich, rise-in-the-world ambitions. Whatever it was that they wanted, or, rather, that they *thought* they wanted, it was not the domestic peace and joy of the self-sufficient farm. Hence Jefferson's attitude is not to be confused with a naïve trust in the fulfillment of the pastoral hope. Recognizing that the ideal society of the middle landscape was unattainable, he kept it in view as a kind of model, a guide to long-range policies as indispensable to intelligent political thought or action as the recognition of present necessities. Like certain great poets who have written in the pastoral mode, Jefferson's

genius lay in his capacity to respond to the dream yet
to disengage himself from it. To a degree, he exemplifies
the kind of intelligence which Keats thought characteristic
of men of achievement, especially in literature, and which
Shakespeare possessed above all others, that is, "negative
capability": the capacity "of being in uncertainties, mysteries, doubts, without any irritable reaching after fact
and reason. . . ." Though Jefferson certainly did reach
after fact and reason, he was able to function brilliantly
under stress of doubt, and his superb mind maintained
its poise as it moved ceaselessly between real and imagined
worlds. As indicated by the example of Shakespeare, creator of *The Tempest* and *King Lear*, an exultant pastoral
wish-image may well emanate from a mind susceptible to
the darkest forebodings. Looking to America's future,
Jefferson anticipates the tragic ambivalence that is the
hallmark of our most resonant pastoral fables. "Our
enemy," he writes during the War of 1812, "has indeed
the consolation of Satan on removing our first parents
from Paradise: from a peaceable and agricultural nation,
he makes us a military and manufacturing one." [51]

The Machine

Locke sank into a swoon;
The Garden died;
God took the spinning-jenny
Out of his side.

   W. B. Yeats, *The Tower*, 1928 *

. . . the most idealist nations invent most machines. America
simply teems with mechanical inventions, because nobody in
America ever wants to *do* anything. They are idealists. Let
a machine do the thing.

D. H. Lawrence, *Studies in Classic American Literature*, 1923

In his *Life of Johnson* Boswell describes the
moment when he first recognized the significance of the
new technology. In March 1776, he visited Matthew Boulton's great Soho works where a steam engine was in production. The "very ingenious proprietor" himself was his
guide. "I wish," Boswell writes, "that Johnson had been
with us: for it was a scene which I should have been
glad to contemplate by his light. The vastness and the
contrivance of some of the machinery would have 'matched
his mighty mind.' " In Johnson's absence, however, Boswell quotes the *"iron chieftain's"* own account of his work.
"I shall never forget Mr. Bolton's [sic] expression to me:
'I sell here, Sir, what all the world desires to have —
POWER.' "

As Boswell tells the story, there is no trace in it of that

---

* Reprinted by permission of Mrs. W. B. Yeats, The Macmillan Co., New York,
and Messrs. Macmillan and Co., Ltd., London.

contempt for the machine later to become a stock literary attitude. In many ways, in fact, he anticipates the dominant American response. He finds the spectacle exhilarating, and his notion that the growing power of technology somehow is a "match" for the power of intellect implies a progressive idea of history. At the time most Americans no doubt shared this view. On the other hand, they surely did not assume, as Boulton does, that machine power entails manufacturing on a large scale. According to Boswell, there were seven hundred workers on the job at Soho when he paid his visit — a prospect that would have appalled Thomas Jefferson.[1]

1

It did not occur to Jefferson that the factory system was a necessary feature of technological progress. In 1786, the year after the first printing of *Notes on Virginia*, with its plea that America let its workshops remain in Europe, Jefferson was in England. This was a moment, as Boulton put it in a letter to his collaborator, James Watt, when the population seemed to have gone "steam-mill mad." At Blackfriar's Bridge, near London, there was a new mill powered by Boulton and Watt engines which was generally considered one of the mechanical wonders of the age, and Jefferson went to see it. He was delighted. "I could write you volumes," he said in a letter to Charles Thomson afterward, "on the improvements . . . made and making here in the arts." Of course Jefferson's passion for utilitarian improvement, gadgets, and labor-saving devices of all kinds is familiar to anyone who has read his letters or visited Monticello. But in England at this time the new technology was visibly related to the new factory system, and one therefore might have expected to hear Jefferson sound another, less enthusiastic, note. But not so. He singles out the steam mill as deserving of particular notice because, he says, it is "simple, great, and likely to have extensive consequences." And he is not thinking about consequences only in England.

> I hear you are applying this same agent in America to navigate boats, and I have little doubt but that it will be applied generally to machines, so as to supercede the use of water ponds, and of course to lay open all the streams for navigation. We know that steam is one of the most powerful engines we can employ; and in America fuel is abundant.[2]

Today Jefferson's attitude is bound to seem curious. Why, we cannot help asking, does he fail to connect the new machinery with Soho and the transformation of England into a vast workshop? Why does he want the latest, most powerful machines imported to America if he would have factories and cities kept in Europe?

Part of the answer is that Jefferson's attitude reflects American economic realities. At the time industrialization scarcely had begun in America. Native manufactures were primitive. The war had stimulated production, to be sure, but chiefly in the form of household industry. In America there was nothing comparable to Soho, nor would there be for a long time. True, ritual gestures toward the "promotion" of manufactures often had been made, notably by Benjamin Franklin, but even he assumed that industry, as compared with agriculture, would be of trivial significance. All reliable opinion supported this view. In *The Wealth of Nations* (1776), the work of political economy to which the age deferred beyond all others, Adam

Smith had warned Americans that it would be folly to direct capital into manufactures. Everyone repeated his sensible argument. During the war, especially, the British delighted in reminding their difficult cousins that, even if they won political independence, they could count upon protracted economic subservience. Edmund Burke popularized this idea, and the Earl of Sheffield summarized it with an incontrovertible body of facts and figures in his *Observations on the Commerce of the United States* (1783). An advocate of a tough policy toward America, Sheffield demonstrated the Republic's helplessness. No matter how oppressive a policy the British followed, he said, there was no danger of provoking serious competition from America. The book was widely read, going through six editions by 1791, and it helped to establish the idea that America's economic development would be unusual, if not unique. Not that many people were disposed to quarrel with Sheffield. Most American statesmen accepted his seemingly flawless case. "I say," John Adams had written to Franklin in 1780, "that America will not make manufactures enough for her own consumption these thousand years." [3]

Under the circumstances, there was nothing farfetched about the prophecy. What made it plausible was not so much the absence of factories and machines, it was the geo-political situation of the country. One has only to recall how small an area had been occupied by 1786; most of the continent was unsettled and, for that matter, unexplored; nine out of ten Americans lived on farms; land was cheap, if not free, and capital was scarce. To be sure, political independence had removed Parliament's legal prohibitions against colonial manufactures, but it is doubtful whether they had made much difference. Geography had been more effective than any laws could have been in blocking manufactures, and it still seemed to present overwhelming obstacles. The most formidable was simply the presence of the land itself — that "immensity of land courting the industry of the husbandman" which encouraged Jefferson in the hope that (the war being over) his countrymen would gladly revert to their former dependence on England. The availability of land worked against manufactures in two ways: it provided an inducement to agriculture, and it dispersed the people over an area too large to be a satisfactory market for manufactures. It is not enough, therefore, to think of the landscape at this time merely as an emblem of agrarian sentiments; it is a perpetual reminder of American differences, a visual token of circumstances which guarantee that the nation's workshops will remain in Europe. But workshops, to Jefferson, are one thing and machines another. It is no accident that his enthusiasm is aroused by a mechanized grist mill — a piece of machinery peculiarly suited to a rural society.* [4]

Quite apart from the state of the American economy, however, there are compelling reasons for Jefferson's failure to see the new machines as a threat to his rural ideal. For one thing, the very notion of "technology" as an agent of change scarcely existed. (It was not until 1829 that Jacob Bigelow, a Harvard professor, coined the word itself.) Although many features of what we now call indus-

---

* Ironically, the American inventor, Oliver Evans, already was perfecting a more impressive version back in the United States. Indeed, his water-powered mill, which was in operation by 1787, is now considered the world's first automatic production line. By using both horizontal and vertical conveyor belts, Evans was able to eliminate all human labor between loading the grain at one end of the continuous process and covering the barrels filled with flour at the other. Later he included even the final packing in the automated process.

trialism already were visible, neither the word nor the concept of a totally new way of life was available. Today our view of history is so deeply colored by an appreciation, if not awe, of technology as an agent of change that it is not easy to imagine Jefferson's state of mind as he inspects the powerful engines at Blackfriar's Bridge. Curiously enough, his very devotion to the principles of the Enlightenment obscures his perception of causal relations we now take for granted. Assuming that knowledge inescapably is power for good, he cannot imagine that a genuine advance in science or the arts, such as the new steam engine, could entail consequences as deplorable as factory cities.[5]

From Jefferson's perspective, the machine is a token of that liberation of the human spirit to be realized by the young American Republic; the factory system, on the other hand, is but feudal oppression in a slightly modified form. Once the machine is removed from the dark, crowded, grimy cities of Europe, he assumes that it will blend harmoniously into the open countryside of his native land. He envisages it turning millwheels, moving ships up rivers, and, all in all, helping to transform a wilderness into a society of the middle landscape. At bottom it is the intensity of his belief in the land, as a locus of both economic and moral value, which prevents him from seeing what the machine portends for America.

2

By this time, however, there were some Americans — not many — whose predilections enabled them to foresee what Jefferson could not. The most astute was Tench Coxe, an ambitious, young Philadelphia merchant who was disturbed by the unhappy state into which American affairs had fallen by the end of the Revolutionary War. Although only thirty-one, Coxe already had made a name for himself as a capable spokesman for the nascent manufacturing interest. Later he was to be Alexander Hamilton's assistant in the Treasury, where he would play an important part in drawing up the *Report on Manufactures* of 1791. If Coxe has been slighted by historians, it is partly because he is an unattractive figure. After the Revolution he was accused of collaboration with the enemy, though he was never brought to trial. Hard-pressed financially, he later maneuvered endlessly for government posts. John Quincy Adams called him a "wily, winding, subtle, and insidious character." This judgment may be somewhat extreme, but it seems clear that Coxe was a careerist and political trimmer whose views seldom were far removed from the interests of his class. Except on the subject of economic development, where he was far ahead of his time, Coxe did not display intellectual distinction. His values were pragmatic in the limited material sense. Yet it was precisely the narrow, prudential quality of his mind that made him responsive to the forces which were to determine the main line of national development. In this sense at least he is entitled to the tribute of Joseph Dorfman, who considers him "closest to being the Defoe of America." At any rate, his claim to our attention rests on two talents: first, a rare empirical bent which led him to make predictions based chiefly upon economic data collected and interpreted by himself, and, second, a master publicist's knack of casting his own aims in the idiom of the dominant ideology. With a sure sense that machine technology would make a decisive difference in the nation's development, Coxe proceeded to make a place for it in the myth of the garden.[6]

In September 1786, just a few months after Jefferson's visit to the mill at Blackfriar's Bridge, Coxe attended the Annapolis meeting on commercial regulations which in turn led to the calling of the Constitutional Convention of 1787. By the following summer his anxiety about the deterioration of the economy of the young Republic had deepened. He was convinced that nothing less than the "salvation of the country" rested upon the delegates who were assembling to write, as it turned out, the Constitution of the United States. Although not a delegate himself, Coxe made his ideas known to the Convention. On May 11, three days before the scheduled meeting, he addressed the Society for Political Enquiries at the home of Benjamin Franklin. Some fifty leading citizens belonged to the organization, which met every two weeks to discuss topics of general interest. Later Coxe saw to it that his speech (*An Enquiry into the principles, on which a commercial system for the United States of America should be founded* . . . [and] *some political observations connected with the subject*) was published and "inscribed to the members of the convention." Then again, on August 9th, while the Convention still was in session, he elaborated upon his theme. At the request of Dr. Benjamin Rush, the president, he gave the inaugural address at the organizing meeting of the Pennsylvania Society for the Encouragement of Manufactures and the Useful Arts. Taken together, these two speeches outline the case for industrialization as a means of realizing the ideal of the middle landscape.[7]

Like most of the official delegates to the Convention, Coxe argues for a stronger, more centralized government with power to enforce uniform economic policies. On the political side, accordingly, he expounds the standard doctrine of the capitalist and nationalist groups. But Coxe is more interested than most in long-range economic goals. Convinced that political independence ultimately will require greater economic self-sufficiency, he insists upon the need for a "balanced economy" and thus, above all, for native manufactures. Without them the young nation's prosperity and its security always will be precarious. At the outset, then, he runs head on into the whole body of respectable economic theory that denies the feasibility of American manufactures. His problem, he admits, is to "disencumber" the case for manufactures of the usual objections: *"the high rate of labour, . . . the want of a sufficient number of hands . . . , — the scarcity and dearness of raw materials — want of skill* in the business itself and *its unfavorable effects on the health of the people."* Impressive as they are, all of these objections may be disposed of, according to Coxe, by one new fact under the sun. It is so new that he does not even have a name for it, but he is certain that it will make all the difference, and so he says:

> Factories which can be carried on by water-mills, windmills, fire, horses and machines ingeniously contrived, are not burdened with any heavy expense of boarding, lodging, clothing and paying workmen, and they supply the force of hands to a great extent without taking our people from agriculture. By wind and water machines we can make pig and bar iron, nail rods, tire, sheet-iron, sheet-copper, sheet-brass, anchors, meal of all kinds, gunpowder . . .

And so on. Describing the incredible productive power of machines and factories, Coxe becomes excited; a note of wonder and prophecy gets into his sober discourse:

Strange as it may appear they also card, spin and even weave, it is said, by water in the European factories.

Steam mills have not yet been adopted in America, but we shall probably see them after a short time. . . .

Machines ingeniously constructed, will give us immense assistance. — The cotton and silk manufacturers in Europe are possessed of some, that are invaluable to them. Several instances have been ascertained, in which a few hundreds of women and children perform the work of thousands of carders, spinners and winders.

The cumulative effect of all this is to undermine most prevailing expectations about the future American economy. Once the new technology is brought into the picture, Coxe is saying, the prospect changes completely. Indeed — and this is the crux no one else seems to have recognized — the very factors usually cited as inhibiting to the nation's growth, such as the scarcity of labor, then will become stimulants. Paradoxically, the extraordinary abundance of land in America is what lends a unique significance to the machine. Coxe knows that the practical, conservative men in his audience will suspect him of being a "visionary enthusiast," but he cannot restrain himself. So inspired is he by the changes to be wrought by machines that he is not satisfied merely to reject the idea of the young Republic's subservience to Europe:

. . . combinations of machines with fire and water have already accomplished much more than was formerly expected from them by the most visionary enthusiast on the subject. Perhaps I may be too sanguine, but they appear to me fraught with immense advantages to us, and not a little dangerous to the manufacturing nations of Europe; for should they continue to use and improve them, as they have heretofore done, their people may be driven to us for want of employment, and if, on the other hand, they should return to manual labour, we may underwork them by these invaluable engines.[8]

To appreciate Coxe's prescience it is necessary to recall how conjectural these ideas were in the summer of 1787. Although machine production was becoming an accepted fact of life in England, it was little more than an idea in America. But this is not to suggest that Coxe was the only American drawn to the possibility. In the small but influential group of "friends of American manufactures" who made up the new Philadelphia society there were a number of men who had expressed one or the other of Coxe's thoughts on the subject. In 1787, moreover, Matthew Carey had founded a new journal, *The American Museum,* which took the same general line. Besides, it is misleading to stress the backwardness of native manufactures. Beneath the surface of economic life the colonies had accumulated a rich fund of technical knowledge and skill, soon to be revealed in the achievements of inventors like Evans, Fitch, Whitney, and Fulton. When all of these allowances have been made, however, the fact remains that it was Coxe who first gathered these scattered impulses and ideas into a prophetic vision of machine technology as the fulcrum of national power.[9]

But for Coxe the machine is the instrument and not in itself the true source of America's future power. If anything, he exaggerates its European identity, and he frankly advocates certain devious methods of wresting the secrets of technological power from the Old World. At the time the British were attempting, by strict regulations, to prevent any plans or new engines of production or skilled mechanics from leaving the country. As countermeasures,

Coxe and his colleagues advertised for technicians; they offered special bonuses to those who would emigrate — a policy notably successful in the well-known case of Samuel Slater; they also attempted to smuggle machines packed in false crates out of England, and the British foreign service was alerted to intercept these illicit cargoes; for a time, in fact, the Americans and the British played an elaborate game of technological espionage and counterespionage. Although Coxe led the campaign to import the new methods, he was intelligent enough not to conceive of American power as emerging from technology *per se*, but rather from the peculiar affinities between the machine and the New World setting in its entirety: geographical, political, social, and, in our sense of the word, cultural.[10]

In his 1787 speeches Coxe displays a striking sensitivity to the prevalence of rural pieties. He tactfully defers to agriculture as the "great leading interest" and repeatedly insists upon the subordinate rôle of manufactures. By employing workers unsuited to farming and thereby helping to develop a home market, he argues, the factory system will benefit agriculture. He anticipates all of the stock protectionist arguments of the next century. Farmers and planters, he says, are the "bulwark of the nation," and their pre-eminence will grow with the "settlement of our waste lands." In his references to the wilderness, incidentally, there is not a trace of primitivism or literary sentimentality. Unimproved land, he says, is "vacant" — a "waste." When discussing the middle landscape, moreover, he adopts the familiar pastoral attitudes. Of course, he says, "rural life promotes health and morality by its active nature, and by keeping our people from the luxuries and vices of the towns." To allay the fears of Jeffersonians, he relies chiefly upon the "safety valve" argument. After all, "the states are possessed of millions of vacant acres . . . that court the cultivator's hand," so how can anyone question the "great superiority of agriculture over all the rest [of the economic interests] combined"? It is impossible to tell how much of this is the calculated strategy of a "gladiator of the quill" — as William Maclay referred to Coxe — and how much he really believed.[11]

At any rate, instead of denying that the economic supremacy of agriculture ultimately inheres in the close relation between farmers and the soil (as Hamilton is tempted to do in the *Report on Manufactures* four years later), Coxe ingeniously contrives an equivalent argument for manufacturing. Building factories, he claims, is necessary to fulfill imperatives embedded in the terrain. "Unless business of this kind is carried on," he says, "certain great *natural powers* of the country will remain inactive and useless. Our numerous mill seats . . . would be given by Providence in vain." (Throughout Coxe is at pains to break down the association of manufactures with the "artificial" as against the alleged "naturalness" of a rural life.) Describing the policy to be inferred from the presence of numerous mill seats, he continues:

> If properly improved, they will save us an immense expence for the wages, provisions, cloathing and lodging of workmen, without diverting the people from their farms — Fire, as well as water, affords . . . a fund of assistance, that cannot lie unused without an evident neglect of our best interests. Breweries, . . . distilleries, . . . casting and steel furnaces . . . are carried on by this powerful element. . . . 'Tis probable also that a frequent use of steam engines will add greatly to this class of factories.

Nowhere is Coxe's genius as a propagandist more evident than in the way he depicts the aims of American society as emanating from geography. He presents his program of economic development as part of a grand topographical design. Not only the abundance of resources but the breadth of the ocean supports his case for developing native manufactures. In fact, America is better suited to the purpose than Europe. With the aplomb of a public relations expert, he turns the standard symbols of the pastoral myth to his own uses. The "clear air and powerful sun of America" will give producers of linens and cottons a distinct advantage over their overseas competitors when it comes to bleaching because, he says, the "European process by drugs and machines impairs the strength." (Although this apparent slur on machines may seem inconsistent, it fits the Coxean formula — the notion that machines merely bring out powers latent in the environment. Thus textile production is more "naturally" suited to America than to Europe.) As for the alleged immorality and ill-health of factory workers, Americans need not be concerned about that. He says that the objection to manufactures as "unfavourable to the health" is urged principally against the production of textiles which *"formerly* were entirely manual and sedentary occupations." The use of machine power erases that old objection.

At this point Coxe anticipates what was to become a central theme in the ideology of American industrialism: the capacity of the New World environment to "purify" the system. Just as the American sun is a more potent bleaching agent, so the entire social climate of the new Republic will cleanse the factory system of its unfortunate feudal residues. Later this idea also would ease Jefferson's mind: in 1805, recalling his diatribe against manufactures in Query XIX, he explains to a correspondent that in the 1780's he had been thinking about workers "of the great

cities in the old countries . . . with whom the want of
food and clothing . . . [had] begotten a depravity of
morals, a dependence and corruption" he had no wish to
see repeated in America. But now, he observes, American
"manufacturers are as much at their ease, as independent
and moral as our agricultural inhabitants, and they will
continue so as long as there are vacant lands for them to
resort to. . . ." Later variants of this refrain would be
heard from European visitors, many of whom would
develop the contrast between the pure, apple-cheeked farm
girls in American mills (the "'nuns of Lowell" in Chevalier's resonant phrase) and their pathetic European
counterparts. Ironically, the sentiment rests at bottom
upon the idea that the factory system, when transferred to
America, is redeemed by contact with "nature" and the
rural way of life it is destined to supplant.[12]

But to return to Tench Coxe in 1787. In addition to
his foresight and the subtlety with which he adapts his
program to rural values, what is most impressive about
his thought is his responsiveness to the topographical and,
indeed, "mythic" quality of the dominant ethos. He fully
appreciates the function of the landscape as a master image
embodying American hopes. At a decisive point in his
first speech, accordingly, he offers by way of summary "to
draw a picture of our country, as it would really exist
under the operation of a system of national laws formed
upon these principles." What follows is a geo-political
landscape painting:

> In *the foreground* we should find the mass of our citi-
> zens — the cultivators (and what is happily for us in
> most instances the same thing) the independent pro-
> prietors of the soil. Every wheel would appear in motion
> that could carry forward the interests of this great body

> of our people, and bring into action the inherent powers of the country. . . . *On one side* we should see our manufactures encouraging the tillers of the earth. . . . Commerce, *on the other hand* . . . would come forward with offers to range through foreign climates in search of . . . supplies . . . which nature has not given us at home. . . .

Coxe has no difficulty blending factories and machines into the rural scene. Combining the best of art with the best of nature, the picture matches the pastoral ideal of the middle landscape.

As a frame for his whole program, Coxe begins his second speech of 1787 by invoking the American moral geography:

> Providence has bestowed upon the United States of America means of happiness, as great and numerous, as are enjoyed by any country in the world. A soil fruitful and diversified — a healthful climate — mighty rivers and adjacent seas abounding with fish are the great advantages for which we are indebted to a beneficent creator. Agriculture, manufactures and commerce, naturally arising from these sources, afford to our industrious citizens certain subsistence and innumerable opportunities of acquiring wealth.

In arguing for the development of machine power, Coxe depicts it as "naturally arising," like agriculture, from the divine purpose invested in the New World landscape. To Matthew Boulton steam engines represented simple, stark power, but Coxe understands that it is wise to represent the machine to Americans as another natural "means of happiness" decreed by the Creator in his design of the continent. So far from conceding that there might be anything alien or "artificial" about mechanization, he insists that it is inherent in "nature," both geographic and human. On the "subject of mechanism," he says, "America may justly pride herself. Every combination of machinery may be expected from a country, A NATIVE SON of which, reaching this inestimable object at its highest point, has epitomized the motions of the spheres, that roll throughout the universe."

With this deft allusion to David Rittenhouse and his orrery, Coxe brings his celebration of the machine in its New World setting to a climax. By reminding his audience of the famous orrery, a miniature planetarium which had won the Pennsylvania scientist international fame, he enlists the immense prestige of Newtonian mechanics in support of his economic program. The orrery is an ingenious replica of the universe: when the clockwork machinery turns, the heavenly bodies revolve in their orbits, music plays, and dials move indicating the hour, the day of the month, and the year. Here, Coxe implies, is a visual and auditory display of the same harmonious plan which has provided America with endless resources for manufactures. If a colonial farmer's son can "epitomize" the ultimate laws of nature — the very music of the spheres — then imagine what Americans will accomplish when they apply the same principles to their entire national enterprise! * At this point the impressive, Miltonic reach and grandeur of Coxe's rhetoric imparts a metaphysical sanction to his vision of American economic development. As he describes the situation in 1787, the momentous achievements of science, the political movement to establish the new American Republic and the forthcoming use of

---

* As if sharing Coxe's sentiments, incidentally, the new Society for the Encouragement of Manufactures and the Useful Arts proceeded to elect Rittenhouse to its vice-presidency.

machine power in production all belong to the same encouraging flow of history. They are all signs of a progressive unfolding of the structural principles of the universe — the laws of "mechanism."

There are few words whose shifting connotations register the revolution in thought and feeling we call the "romantic movement" more clearly than "mechanism." Once the influence of Wordsworth, Coleridge, and Carlyle had been felt in America, no writer (whether sympathetic with them or not) would find it possible to use "mechanism" in the unself-conscious, honorific sense in which Coxe uses it. His entire argument for what we would call industrialization rests on the assumption that celestial mechanics, the orrery, the new engines of production, even the factory system — all embody the same ultimate laws of nature. What is more, and this is perhaps the most difficult attitude to grasp in retrospect, it is the same "mechanism" to which we respond, aesthetically, in the presence of the natural landscape. The identification of visual nature with the celestial "machine" is difficult to grasp because of our own feeling, learned from the romantics, that "organic" nature is the opposite of things "mechanical." But it is impossible to appreciate the dominant American attitude toward technology if we project this sense of contradiction too far back into the past. In other words, Coxe, who anticipates the popular view, is writing in the tradition of James Thomson, whose feeling for the beauty of the countryside was inseparable from his reverence for the Newtonian world machine. On the occasion of Sir Isaac's death, Thomson had described the great man in these terms:

> All intellectual eye, our solar round
> First gazing through, he by the blended power

> Of gravitation and projection, saw
> The whole in silent harmony revolve.

And when Newton studied the stars they

> . . . Blazed into suns, the living center each
> Of an harmonious system: all combin'd
> And rul'd unerring by that single power,
> Which draws the stone projected to the ground.

Thomson's version of the Newtonian system is one of innumerable examples, a variant of the metaphor which George Berkeley had used in his *A Treatise Concerning the Principles of Human Knowledge* (1710):

> Such is the artificial contrivance of this mighty machine of nature that, whilst its motions and various phenomena strike on our senses, the hand which actuates the whole is itself unperceivable to men of flesh and blood. . . .

True, the "mighty machine" can be described with most precision in the language of mathematics; but this does not prevent the poets of the age, particularly those who celebrate the glories of the rural scene, from communicating a sense of Newton's universe in metaphor. Thus William Somerville toward the end of "The Chase" (1735):

> . . . grand machine,
> Worlds above worlds; subservient to his voice,
> Who, veil'd in clouded majesty, alone
> Gives light to all; bids the great system move,
> And changeful seasons in their turns advance,
> Unmov'd, unchang'd, himself . . .[13]

The speeches of Tench Coxe in the summer of 1787 prefigure the emergence of the machine as an American cultural symbol, that is, a token of meaning and value recognized by a large part of the population. By 1851, when Walt Whitman tells the Brooklyn Art Union that the United States has become a nation "of whom the steam engine is no bad symbol," he assumes that his audience knows what he is talking about. Needless to say, a collective image of this kind gathers meanings gradually, over a long period, and it is impossible to fix upon any single moment when it comes into being. Besides, it invariably combines a traditional meaning and a new, specific, local or topical reference. The garden image brings together a universal Edenic myth and a particular set of American goals and aspirations. So with the machine. What is most fascinating about the speeches of Tench Coxe is that in them we witness the virtual discovery of the symbolic properties of the machine image — its capacity to embrace a whole spectrum of meanings ranging from a specific class of objects at one end to an abstract metaphor of value at the other.[14]

To Coxe the new and most exciting implication of "machine" is technological. In the summer of 1787 the possibilities of the latest technological innovations were just beginning to take hold of the American imagination. While the Constitutional Convention was in session, as it happens, "Mad" John Fitch managed, after inconceivable trials and tribulations, to propel a steam-powered vessel against the current of the Delaware River. When some of the delegates went down to the waterfront to inspect this remarkable invention, what they saw was another instance of, in Coxe's language, the power of mechanism. The existence of this kind of machine leads him to an optimistic view of the Republic's future, and there can be no doubt that in his mind the development of steam power and the business of the Constitutional Convention are aspects of the same grand enterprise. They both represent a release of power through seizure of the underlying principles of nature. The universe is a "mechanism." At the abstract end of the spectrum, then, the symbol of the machine incorporates a whole metaphysical system. It often has been noted that the dominant structural metaphor of the Constitution is that of a self-regulating machine, like the orrery or the steam engine; it establishes a system of "checks and balances" among three distinct, yet delicately synchronized, branches of government.[15]

Between the two extremes, the machine as concrete object and the machine as root metaphor of being, Coxe identifies the power of "mechanism" with a specific economic faction. He is an avowed spokesman for the manufacturing interest. He and his associates in the Pennsylvania delegation are chiefly concerned about the kind of power to be generated by the national government's authority to enforce uniform economic laws, protect patents, and establish tariffs. Running through much of what they had to say is an inchoate sense of the vast transformation of life to be accomplished through what we should call economic development or industrialization. "The time is not distant," said Gouverneur Morris at one point in the debates, "when this Country will abound with mechanics & manufacturers who will receive their bread from their employers." But this is a rare example; the vocabulary at the command of Coxe, Morris, and their friends was inadequate to express their full sense of the power to be released through the combined agencies of commerce, science, technology, and republican institutions. To speak of it as "manufacturing" power hardly sufficed, particularly in view of the fact that the word still carried much of its pre-industrial meaning, as in handicraft or household manufactures. Their vision had got ahead of their language.* [16]

Still, we can be sure about the widespread identification of the manufacturing interest, the new Republic, and the new technology. On July 4, 1788, after the Constitution had been approved by the electorate, there were parades and pageants in many cities, and several featured displays of recent industrial progress. In Philadelphia, the new "manufacturing society" entered a float in the grand Federal Procession. It was thirty feet long and thirteen feet wide, and drawn by ten large bay horses. On board eleven men and women demonstrated the operation of the latest machinery used in textile manufacture — a spinning jenny, a carding machine, and a loom. A new American flag flew above the float, and to it was attached the motto: "May the Union Government Protect the Manufacturers of America." One hundred weavers marched behind the float carrying a banner inscribed, "May Government Protect Us." Half a century later, when industrialization was well under way, the notion that 1789 marked the beginning of the process became something of a commonplace. The chief evil of the colonial situation, said Edward Everett in 1831, had been the "restraint" upon "the

* Actually, a whole new vocabulary of "industrialism" was just emerging at this time. In the 1780's and '90's words like "engine," "machine," "manufacture" and "industry," all of which had been in use long before the advent of power machinery, were beginning to take on new meanings. At one point Coxe refers to "handicraft manufactures," indicating that for him at least the word now had become ambiguous. (*A View of the United States,* p. 38n.) On the other hand, David Humphreys published "A Poem on the Industry of the United States of America" in 1802, in which he still used "industry" in its traditional sense as a character trait (diligence) rather than as an economic institution. A careful study of this shift in language would be invaluable for a detailed understanding of the impact of industrialization upon consciousness.

labor of the country," and "the first thought and effort of our fathers . . . [was to] encourage and protect the mechanical arts and manufactures. . . ." [17]

Among other things, Everett had in mind the *Report on the Subject of Manufactures* which Alexander Hamilton, assisted by Tench Coxe, prepared for the Congress in 1791. This deservedly famous state paper reflects the unmistakable shift in attitudes toward manufactures that coincided with the formation of the new government. Only four years had passed since Coxe had expounded his ideas in Philadelphia, but Hamilton assumed that they had been widely accepted. "The expediency of encouraging manufactures in the United States," he says in the first sentence, "which was not long since deemed very questionable, appears at this time to be pretty generally admitted." [18]

In many ways the *Report* goes far beyond Coxe's speeches; it is composed with a systematic rigor and comprehensiveness that makes his thought seem crude and impressionistic. But there is one respect in which Coxe's argument had been more subtle: it took into account, as the *Report* does not, the hold of the pastoral ideal upon the national consciousness. Although Hamilton begins with a routine effort to placate the agricultural interest, he is an undisguised advocate of continuing economic development. He offers no equivalent of Coxe's symbolic landscape, or his vague implication that, somehow, technology would help America reach a kind of pastoral stasis. In fact, he finally makes no rhetorical concessions to Jeffersonian hopes and fears. To support his argument for machine production, he describes the advantages of the "cotton-mill, invented in England, within the last twenty years . . ." in this blunt language:

. . . all the different processes for spinning cotton, are performed by . . . machines, which are put in motion by water, and attended chiefly by women and children; and by a smaller number of persons, in the whole, than are requisite in the ordinary mode of spinning. And it is an advantage of great moment, that the operations of this mill continue with convenience, during the night as well as the day. The prodigious effect of such a machine is easily conceived. To this invention is to be attributed, essentially, the immense progress which has been so suddenly made in Great Britain, in the various fabrics of cotton.[19]

Whether he fully intends to endorse it or not, Hamilton makes no effort to disclaim the chilling idea of putting women and children on the night shift in cotton mills. The passage exemplifies the tough, hard-boiled tone of the *Report*. Throughout Hamilton is forthright, unsentimental, logical, and clear. Taken as a whole, in fact, the *Report* is a blueprint for a society aimed at maximum productivity, not as an end in itself, but as the key to national wealth, self-sufficiency, and power. The power of the United States as a corporate entity is the ultimate goal; what Hamilton wants is the economy best suited to the establishment of America's supremacy among nations. There can be little doubt that Tench Coxe wanted much the same sort of society. Unlike Hamilton, however, Coxe saw the need to couch this aim in the language of the prevailing ideology. No matter what the economic behavior of his countrymen might seem to indicate, Coxe understood that they preferred not to acknowledge wealth and power as their goals. In this sense he was a subtler and more farsighted — if less candid — advocate of industrialization than Hamilton. He foresaw that Americans would be more likely to endorse the Hamiltonian program with enthusiasm if permitted to conceive of it as a means of fulfilling the pastoral ideal.

3

By 1830 a contrasting image of the machine had begun to attract the attention of American intellectuals. It came out of Germany by way of England. Friedrich Schiller, in his *Letters upon the Aesthetical Education of Man* (1795), had brought in a "complicated machine," "the monotonous sound of the perpetually revolving wheel," to figure the "degeneration" of contemporary culture. What was then happening in Europe, according to Schiller, resembled the decline of classical Greek culture. A "common and coarse mechanism" was setting in, society changing from one in which the individual enjoyed — or at least might have enjoyed — an "independent life" enabling him to become "a separate whole and unit in himself," into one dominated by "an ingenious mechanism." The state itself was "splitting up into numberless parts" and so was mankind:

> Man himself, eternally chained down to a little fragment of the whole, only forms a kind of fragment; having nothing in his ears but the monotonous sound of the perpetually revolving wheel, he never develops the harmony of his being; and instead of imprinting the seal of humanity on his being, he ends by being nothing more than the living impress of the craft to which he devotes himself, of the science that he cultivates. This very partial and paltry relation, linking the isolated members to the whole, does not depend on forms that are given spontaneously; for how could a complicated

machine, which shuns the light, confide itself to the free will of man?

Here Schiller is using "machine," in the technological sense, to represent a "mechanistic" social system, the increasingly complex kind of society emerging along with the new machine power.[20]

Of the English writers who used this device, none was to have a greater influence in America than Thomas Carlyle. He had published his "Life" of Schiller in 1823-4, and in his *Edinburgh Review* essay "Signs of the Times" (1829), he seizes upon the machine as the most telling "sign" of modern life. If asked to select a name for the age, he says, he would not call it the "Heroical, Devotional, Philosophical, or Moral Age, but, above all others, the Mechanical Age. It is the Age of Machinery, in every outward and inward sense of that word . . ." Playing upon all possible connotations of "machinery," Carlyle turns it into the controlling symbol for a new kind of culture. It is the culture, the inner world of thought and feeling, that really interests him, but he regards the image of machinery as representing the causal nexus between the new culture and the outer world, or society. Accordingly, he uses the words "machine," "machinery," and "mechanism" in two distinct, though finally complementary, ways.[21]

First, there is the "outward" sense: "machine" points directly to the new technology:

> Our old modes of exertion are all discredited, and thrown aside. On every hand, the living artisan is driven from his workshop, to make room for a speedier, inanimate one. The shuttle drops from the fingers of the weaver, and falls into iron fingers that ply it faster. . . . Men have crossed oceans by steam; the Birmingham

Fire-king has visited the fabulous East. . . . There is no end to machinery. Even the horse is stripped of his harness, and finds a fleet fire-horse yoked in his stead. . . . For all earthly, and for some unearthly purposes, we have machines and mechanic furtherances; for mincing our cabbages; for casting us into magnetic sleep. We remove mountains, and make seas our smooth highway; nothing can resist us. We war with rude Nature; and, by our resistless engines, come off always victorious, and loaded with spoils.

Carlyle then extends the effects of "machinery" (still in the outward sense) to the emergent social and economic system which he would soon name "industrialism."

. . . What changes, too, this addition of power is introducing into the Social System; how wealth has more and more increased, and at the same time gathered itself more and more into masses, strangely altering the old relations, and increasing the distance between the rich and the poor. . . .

It is the second, or "inward," sense of the word "machine," however, to which Carlyle devotes most attention. What concerns him is the way the "mechanical genius . . . has diffused itself into quite other provinces. Not the external and physical alone is now managed by machinery, but the internal and spiritual also." Here "machinery" stands for a principle, or perspective, or system of value which Carlyle traces through every department of thought and expression: music, art, literature, science, religion, philosophy, and politics. In each category he detects the same tendency: an excessive emphasis upon means as against ends, a preoccupation with the external arrangement of human affairs as against their inner meaning and consequences. Although he is using the image of the machine metaphorically, he does not lose control of the distinction between fact and metaphor. In discussing the functions of government, for example, he admits that they include much that is essentially routine or mechanical. "We term it indeed, in ordinary language, the Machine of Society, and talk of it as the grand working wheel from which all private machines must derive, or to which they must adapt, their movements." In this context, Carlyle recognizes, machine is a figure of speech, yet the figure seems to contain a prophetic truth.

> Considered merely as a metaphor, all this is well enough; but here, as in so many other cases, the 'foam hardens itself into a shell,' and the shadow we have wantonly evoked stands terrible before us, and will not depart at our bidding.

His point is that the age is increasingly reliant upon "mere political arrangements," and that in politics, as in all else, less and less account is being taken of that which "cannot be treated mechanically." Carlyle's immediate target is utilitarianism, with its emphasis upon the proper structure of institutions. But back of that philosophy he sees the environmentalism of the eighteenth century— the view that, on the whole, external conditions determine the quality of life, hence human suffering can best be attacked by contriving better social machinery. What bothers Carlyle is the easy assumption that, as he puts it, "were the laws, the government, in good order, all were well with us; the rest would care for itself!"

In philosophy this mechanistic spirit is reflected in the still high reputation of John Locke. "His whole doctrine," says Carlyle, "is mechanical, in its aim and origin, in its method and its results." When Locke makes the contents of the mind contingent upon images flowing in upon it from the outside, he reduces thought to what is ultimately a reflex of the world "out there." To account for a man's ideas and values only, or even chiefly, by the circumstances in which he lives is, according to Carlyle, to divest his thought of will, emotion, and creative power. If the mind is a reflex of what is, how can it possibly control circumstances? Control implies the power to compare what is with what may be. To Carlyle the empirical philosophy is negative and quietistic. "By arguing on the 'force of circumstances,' " he says, "we have argued away all force from ourselves; and stand lashed together, uniform in dress and movement, like the rowers of some boundless galley." In its transactions with the world outside, a mind so conceived responds like one cogged wheel turned by another. Used in this way the image of the machine connotes loss of inner freedom even as it provides outward power. "Practically considered," says Carlyle, "our creed is Fatalism; and, free in hand and foot, we are shackled in heart and soul with far straiter than feudal chains."

One of the remarkable things about "Signs of the Times" is the clarity and cogency with which Carlyle connects the machine as object (a technological fact) and the machine as metaphor (a token of value). In large part his success is due to a tacit recognition of culture as an integrated whole. Like a modern anthropologist, Carlyle is attempting to make statements about an entire way of life, a complex which embraces all the behavior of Englishmen — their physical activities, their work, their institutions, and, above all, their inner lives. In using the machine as a symbol of the age, he is saying that neither the causes nor the consequences of mechanization can be confined to the "outer" or physical world. The onset of machine power, he says, means "a mighty change in our whole manner of existence." This is the insight which would lead him to use the new word "industrialism," and it helps to explain why, from the beginning, the very idea of an industrial society as a unique phenomenon has been tinged by a strong critical animus. The machine represents a change in our whole way of life, Carlyle argues, because "the same habit regulates not our modes of action alone, but our modes of thought and feeling. Men are grown mechanical in head and in heart, as well as in hand."

To a democratic moralist the chief difficulty in Carlyle's argument would be immediately obvious. In taking a critical attitude toward the new technology he seems to deny the working mass of mankind its unique opportunity to escape lives of dull, repetitive labor. This point was particularly striking to a young American, Timothy Walker, who soon attacked the essay in the pages of the *North American Review*. Before considering Walker's reply, however, we must distinguish Carlyle's position as sharply as possible from the aesthete's nostalgic and facile rejection of industrialism with which we now are so familiar. For one thing, Carlyle does not deny the genuine advantages of machine production. He is not a reactionary in the strict sense of calling for a return to an earlier, simpler society. In fact, he ends his essay with an affirmation of the long-range progress of the race. It is, he says, "a well-ascertained fact, that . . . the happiness and greatness of mankind at large have been continually progressive."

> Doubtless this age also is advancing. Its very unrest, its ceaseless activity, its discontent contains matter of promise. Knowledge, education are opening the eyes of the humblest; are increasing the number of thinking minds

without limit. This is as it should be; for not in turning
back, not in resisting, but only in resolutely struggling
forward, does our life consist.

Where then, we may well ask, is the ground for his attack upon the new industrial order? If he accepts the
value of machine technology, if he believes that there is
no turning back, why does he have this animus? In part
the answer lies in his commitment to a traditional humanist idea of a proper balance of human life. To say
that men have grown "mechanical" in head and heart is
to say that they now over-value those aspects of life which
are calculable and manipulatable and, by the same token,
that they neglect the whole sphere of the spontaneous,
the imaginative — all that springs from the inner resources of the psyche: "the primary, unmodified forces and
energies of man, the mysterious springs of Love, and Fear,
and Wonder, of Enthusiasm, Poetry, Religion, all which
have a truly vital and *infinite* character. . . ." Carlyle
calls this neglected province (the antithesis of the mechanical) the dynamical, and his entire argument must finally
be said to rest upon the proposition that "only in the
right coördination of the two, and the vigorous forward-
ing of both, does our true line of action lie."

> Undue cultivation of the inward or Dynamical province
> leads to idle, visionary, impracticable courses, and, espe-
> cially in rude eras, to Superstition and Fanaticism, with
> their long train of baleful and well-known evils. Undue
> cultivation of the outward, again, though less immedi-
> ately prejudicial, and even for the time productive of
> many palpable benefits, must, in the long-run, by de-
> stroying Moral Force, which is the parent of all other
> Force, prove not less certainly, and perhaps still more
> hopelessly, pernicious. This, we take it, is the grand

characteristic of our age. By our skill in Mechanism, it has come to pass, that in the management of external things we excel all other ages; while in whatever respects the pure moral nature, in true dignity of soul and character, we are perhaps inferior to most civilised ages.

Though he is using a novel vocabulary, Carlyle is invoking a traditional morality. To speak of coördinating the mechanical and dynamical provinces of human behavior is, after all, but another way of describing the desire for a proper balance of man's outer and inner, material and psychic, selves. Carlyle is affirming an ideal akin to the "ethic of the middle link" and to the norm implicit in the pastoral image of America. It is a variant of Jefferson's hope that the new Republic would subordinate economic ambition to the "pursuit of happiness." All of which accounts for the fact that in America Carlyle's critique of the new industrial order was so readily adapted to the pastoral idiom. In his view the new power is to be feared, not as inherently evil, but rather as a threat to the necessary balance in the human situation. By placing undue emphasis upon the external environment, the new technology threatens to destroy moral force.

What Carlyle intends by "destruction of moral force" is akin to what soon would be known as "alienation." This seminal idea, then emerging out of the stream of post-Kantian idealism, has dominated the criticism of industrial society ever since. When Carlyle speaks of men having grown "mechanical in head and heart," he means that their behavior is increasingly determined externally, which is to say, by invisible, abstract, social forces unrelated (or alien) to their inward impulses. Hegel had called this state "self-estrangement," thereby implying a conflict between the "social" and the "natural" self. Later Karl Marx was to make a more explicit connection between this inward state and the conditions of life in the new industrial society. He noted that within capitalist relations of production, accompanying the division of labor and mass manufacturing, the working man's product may well become his "enemy." The more he produces, in other words, the more danger there is that the market will be glutted and that he will lose his job. Hence machine technology is instrumental in creating what Marx calls "alienated labor." * Although it is morally neutral, the machine in a capitalist setting helps to transform the worker into a commodity for sale on the labor market. His work takes on a mechanical, meaningless character. It bears little or no relation to his own purposes. The result is the typical psychic set of industrial man which Marx calls alienation. In Erich Fromm's words, the alienated man is one for whom "the world (nature, others, and he himself) remain alien. . . . They stand above and against him as objects, even though they may be objects of his own creation. Alienation is essentially experiencing the world and oneself passively, receptively, as the subject separated from the object." [22]

Describing in 1844 the situation which gives rise to alienation, Marx uses a language strikingly similar to Carlyle's in "Signs of the Times." "The *devaluation* of the human world," he says, "increases in direct relation with the *increase in value* of the world of things." During the

* He first developed the concept in the economic and philosophical manuscripts he wrote in 1844. These early manuscripts, which were not published until 1932, and not translated into English until 1959, establish even more clearly than before the humanist core of the Marxian critique of capitalism.

1840's variants of this idea swept through the intellectual community on both sides of the Atlantic. Three years after Marx's paper on alienation, Emerson wrote these familiar lines:

> Things are in the saddle,
> And ride mankind.
>
> There are two laws discrete,
> Not reconciled, —
> Law for man, and law for thing;
> The last builds town and fleet,
> But it runs wild,
> And doth the man unking.

Years later, in describing the intellectual climate of this period, Emerson stressed the pervasive sense of "detachment." He found it reflected everywhere: in Kant, Goethe's *Faust*, and in the mood generated by the advance of capitalist power. "Instead of the social existence which all shared," he wrote, "was now separation." He saw a connection, moreover, between this feeling of "separation" and the widespread emotional instability of the intellectuals. The young men of the time, he said, were born with knives in their brains.[23]

When Carlyle speaks of the subordination of the "dynamical" to the "mechanical" aspects of life, he anticipates the post-Freudian version of alienation. In a recent book (*Eros and Civilization*), Herbert Marcuse attributes this state of psychic powerlessness to the increasing repression of instinctual drives made necessary by a more and more complicated technological order. To satisfy the imperatives of the mechanized society, men are called upon to endure an intolerable curbing of their spontaneous, erotic, and passional selves. In writing "Signs of the

Times," unfortunately, Carlyle lacked the systematic theory of the mind and the relatively unambiguous vocabulary of depth psychology. It is now obvious that his work, like much of the nineteenth century's criticism of the new industrial order, was vitiated by the inadequacy of his old vocabulary. Unable to describe man's emotional needs in any other terms, writers often fell back upon an outworn collection of religious and moral counters. In "Signs of the Times," however, Carlyle is reaching toward something like the post-Freudian view when he speaks of "mechanism" as stifling the "primary, unmodified forces and energies of man," or again when he sets the machine in opposition to the "mysterious springs of Love." That he appreciated the connection between the typical emotional crises of the age and industrialization becomes most obvious in *Sartor Resartus* (1833–34). This book, which Emerson read just before writing *Nature,* and which Melville read not long before writing *Moby-Dick,* was to have an immense influence in America. Although not written in the pastoral mode, it embodies attitudes which were easily assimilated to American pastoralism. At the climax Professor Teufelsdröckh, a semi-autobiographical figure, is overwhelmed by the meaninglessness of his life. He can establish no contact with a reality beyond himself. Doubtless such depressed states are not peculiar to industrial societies, but Carlyle imparts a distinctively modern aspect to Teufelsdröckh's depression by the image he uses to evoke the moment of total despair:

> To me [says Professor Teufelsdröckh] the Universe was all void of Life, of Purpose, of Volition, even of Hostility: it was one huge, dead, immeasurable Steam-engine, rolling on, in its dead indifference, to grind me limb from limb.[24]

4

By 1829, when "Signs of the Times" appeared in the *Edinburgh Review,* machine power had begun to play the part in American life that Tench Coxe had foreseen in 1787. A profitable factory system was firmly established in New England; new roads and canals and cities were transforming the landscape; on rivers and in ocean harbors the steamboat was proving the superiority of mechanized transport, and outside Baltimore *the* revolutionary machine of the age was being made ready for use: thirteen miles of the first important American railroad, the Baltimore and Ohio, were under construction. As Coxe had foreseen, the conditions once regarded as obstacles to America's economic growth had become stimulants. So far from perpetuating the "backwardness" of the economy, the abundance of land and the scarcity of labor had intensified the demand for machinery. Between 1830 and 1860 the nation was to put down more than 30,000 miles of railroad track, pivot of the transportation revolution which in turn quickened industrialization. By the time Carlyle's essay was read in America the economy was expanding at a remarkable rate, the new technology was proving itself indispensable, and the nation was on the verge of the "take-off" into the era of rapid industrialization.

On the plane of ideas, meanwhile, the views of Tench Coxe had been widely accepted. Apologists for the southern slavery system aside, there was not (nor would there be) any effective opposition to industrialization. This is not to deny that there were impulses to resist. But on the whole they were sporadic, ineffectual impulses; they provoked a number of vivid, symbolic gestures (chiefly of a nostalgic cast), but they did not produce an alternative theory of society capable of enlisting effective political support. By 1829, in fact, Coxe's notion that the aims of the Republic would be realized through the power of machine production was evolving into what can only be called the official American ideology of industrialism: a loosely composed scheme of meaning and value so widely accepted that it seldom required precise formulation. In the written record it appears chiefly as rhetoric in homage to "progress." With this word the age expressed a faith in man's capacity to understand and control history which is now difficult to recapture. And it was the miraculous machinery of the age, beyond all else, which made it obvious that things were getting better all the time. It is not surprising, then, that Carlyle's passionate attack upon the "Age of Machinery" soon provoked an American reply.[25]

Nor is it surprising that the reply appeared in the *North American Review*. Under the editorship of Alexander Hill Everett, brother of Edward, this journal was a vehicle for the rational and, in the main, politically conservative, theologically liberal opinion of the New England Establishment. Its contributors, many of whom were Unitarian ministers or Harvard professors, included George Bancroft, Orville Dewey, Samuel Eliot, Edward Everett, William Prescott, Jared Sparks, and Joseph Story. The young man who answered Carlyle, Timothy Walker, had been well trained for membership in this solid, professional elite. After a proper education at Harvard, where he graduated first in the class of 1826, Walker had taught school under George Bancroft and later had returned to Harvard to prepare for the law with Justice Story. When his reply to Carlyle was published in 1831, he had begun to practice law in Cincinnati.[26]

Walker, casting himself as America's attorney, calls the essay in which he answers Carlyle a "Defence of Mechanical Philosophy." It is a concise, lucid, and well-organized brief on behalf of technological progress. Unlike Coxe or Hamilton, Walker makes no effort to reconcile the new power with the pastoral ideal of economic sufficiency. But then neither does he acknowledge that any change has occurred in the aims of the Republic. He expounds the doctrine of unlimited economic development as if the American people always had been committed to it, as if there always had been a place for the machine in the myth of the garden. Although it contains little, if anything, that is original, Walker's "Defence" makes a fine exhibit of pervasive attitudes toward the new machine power.

At the outset Walker announces that Carlyle has raised a "grave and solemn" question, namely, "whether mankind are advancing or not, in moral and intellectual attainments." Quickly reviewing his adversary's "cheerless conclusions" about the new "Age of Machinery," Walker asserts that he can find in these views nothing but "phantoms." And yet Carlyle's skepticism is profoundly disturbing to him; it irks him because Carlyle manifestly repudiates Walker's faith, and that of the society for which he speaks, as a hollow faith. Incredulity is the dominant note of the opening section: what possible injury can "Mechanism" have done to man? "Some liberties," Walker agrees, ". . . have been taken with Nature by this . . . presumptuous intermeddler," and he describes them with heavy courtroom irony:

> Where she [Nature] denied us rivers, Mechanism has supplied them. Where she left our planet uncomfortably rough, Mechanism has applied the roller. Where her mountains have been found in the way, Mechanism has boldly levelled or cut through them. Even the ocean, by which she thought to have parted her quarrelsome chil-

dren, Mechanism has encouraged them to step across. As if her earth were not good enough for wheels, Mechanism travels it upon iron pathways.

Now where, asks Walker, "is the harm and danger of this?" Since Carlyle had not grounded his critique of mechanization in sentiment for the countryside, the point of Walker's rhetorical question is a bit obscure. But it allows Walker to inject a cherished American image into the record: the machine as a "supplier" of rivers, a "roller" smoothing over what is "uncomfortably rough" in nature. The machine, in short, is an instrument for making what the age calls "improvements." With its help, a waste land can be transformed into a garden. Yet the capacity of technology to transform the landscape, however much it may engage the native imagination, scarcely meets Carlyle's objections — and Walker admits as much. What troubles Carlyle, he says, is not that "Nature will be dethroned, and Art set up in her place. . . ." Or at least, Walker adds, "not exactly this" — an astute reservation indicating that if Carlyle is concerned about the effect of technology upon "Nature," the "Nature" in question is not primarily the physical, or external, nature associated with landscape.

But then precisely what is threatened by Mechanism? It is typical of Walker's side in this familiar debate that he adopts the condescending tone of a no-nonsense empiricist trying valiantly to fathom the mind of a hopeless obscurantist. What worries Carlyle, he says ("if we rightly apprehend his meaning"), is that "mind will become subjected to the laws of matter; that physical science will be built up on the ruins of our spiritual nature; that in our rage for machinery, we shall ourselves become machines." By using these commonplaces Walker intends to convey the spuriousness of Carlyle's argument. (At points he in-

sinuates that Carlyle is a devious exponent of some nameless "Mysticism.") These phrases are stock expressions of
the widespread and largely impotent anxiety generated by
mechanization; no doubt the most popular, closely akin
to the "men-will-become-machines" trope, was the Frank-
enstein fable: the story of the robot that destroys its heart-
less creator. Back of these clichés, as the twentieth century
has discovered, there was a not wholly fanciful premonition of mankind's improving capacity for self-destruction.
But Walker did not take that possibility seriously. As he
phrases the issue, he intends only to expose the melodramatic hollowness of Carlyle's concern, which is (in
Walker's words) whether or not "mechanical ingenuity
is suicidal in its efforts."

To answer the rhetorical question Walker quickly reviews the contribution of technology to cultural evolution.
"In the first ages of the world," he says, "when Mechanism
was not yet known, and human hands were the only instruments, the mind scarcely exhibited even the feeblest
manifestations of its power." (Although Walker is capable
of extending the meaning of "Mechanism" almost in-
definitely, so that it finally threatens to swallow up science
and, for that matter, Being itself, here it chiefly means
technology both as a reservoir of knowledge and as mechanical apparatus.) Without technology there would be
no culture. Only after the "first rudiments of Mechanism
made their appearance" did men have enough time to
think.

> Leisure gave rise to thought, reflection, investigation;
> and these, in turn, produced new inventions and facili-
> ties. Mechanism grew by exercise. Machines became
> more numerous and more complete. The result was a
> still greater abridgment of labor. One could now do

the work of ten. . . . It is needless to follow the deduction farther.

Proceeding from these postulates, Walker lines up the rest of his argument like a mathematical proof. (He was a master of simple linear logic, and in fact he had recently published a geometry textbook.) With perfect assurance he concludes that the "cultivation of the intellect" is "altogether the result of Mechanism, forcing inert matter to toil for man," and, therefore, he looks "with unmixed delight at the triumphant march of Mechanism. So far from enslaving, it has emancipated the mind, in the most glorious sense." What freedom man achieves is the direct result of mechanical inventions. Walker's theory of the technological basis of culture leads him to a simple formula for the advance of civilization. That nation, he says, "will make the greatest intellectual progress, in which the greatest number of labor-saving machines has been devised." In 1831 Walker envisages an Automated Utopia in which "machines are to perform all the drudgery of man, while he is to look on in self-complacent ease."

To lend an ultimate sanction to technology, Walker, like Tench Coxe in 1787, brings in the metaphysical overtones the machine image had carried over from Newtonian science. His cosmos is constructed according to a knowable, rational, and mathematically precise blueprint. "When we attempt to convey an idea of the infinite attributes of the Supreme Being," says Walker, "we point to the stupendous machinery of the universe." A new power machine, then, is not simply a useful implement, it is a kind of totem. This attitude no doubt explains the extreme repugnance that Carlyle's essay arouses in Walker. Since he regards technological progress as evidence that man is gaining access to the divine plan, a kind of gradual revelation,

Carlyle's attack is blasphemy. Technology supplants man's animal functions, thereby making possible the liberation of mind. From Walker's "Defence" one gets the impression that mankind is destined to arrive at a state, like that enjoyed by the ruler of a universe which operates like clockwork, of divine immobility or pure consciousness. "From a ministering servant to matter," Walker explains, "mind has become the powerful lord of matter. Having put myriads of wheels in motion by laws of its own discovering, it rests, like the Omnipotent Mind, of which it is the image, from its work of creation, and pronounces it good."

Although many of Walker's ideas had been anticipated by Tench Coxe, his tone and attitude are quite different. Whereas Coxe had proposed, with some caution, that the machine could change the fortunes of the Republic, Walker discusses technology in the accent of a true believer. He declares "an unfaltering belief in the permanent and continued improvement of the human race" — of which improvement, he says, "no small portion . . . [is] the result of mechanical invention." He invests the machine with the ebullience — the singular, not to say manic, millennialist spirit — which was widespread at the time. At moments he sounds like an inspired, transcendentalist prophet. But Walker fancies himself a realist, and — charming commentary on this era — he draws a fine line between foolhardy optimists who believe that man may arrive at perfection and more sensible people, like himself, who only credit the perfectibility of man. Having made this reservation, he unashamedly joins those who "see Atlantis, Utopia, and the Isles of the Blest, nearer than those who first descried them." He delights in contemplating "these imaginary abodes of pure and happy beings" because, he explains, even though man may not re-create the golden age, such places "are types and shadows of a higher and better condition of human nature, towards which we are surely though slowly tending."

But the most important value that Walker attaches to the image of the machine is political. He regards the new technology as the instrument appointed to fulfill the egalitarian aims of the American people. He expounds the same explosive idea implicit in the irrepressible epithet "industrial revolution." * Coined in France sometime around 1810, the phrase expresses the close kinship between the two new forces, political and technological, which were to threaten the old order everywhere during the next 150 years. The feature common to this double "revolution" was the unprecedented claim of the propertyless, working masses for a fair share of the necessities and perhaps even the felicities of life. What now made the claim reasonable for the first time was science and technology. In America the plausibility of this novel idea was reinforced by natural abundance. So much so, indeed, that Walker honors the egalitarian implications of the new power without a hint of political radicalism. It does not occur to him that a basic change in the structure of society,

---

* In recent years the epithet has been the subject of an endless and often silly scholarly controversy. Economic historians, in particular, have quarreled about its validity as a descriptive term. For them, the issue seems to be whether there was a break, during the eighteenth century, in the economic and technological development of England so severe as to justify the name "revolution." But the whole issue becomes irrelevant once we recognize that we are dealing with a metaphor, and that its immense appeal rests, not on its capacity to describe the actual character of industrialization, but rather on its vivid suggestiveness. It evokes the uniqueness of the new way of life, as experienced, and, most important, it is a vivid expression of the affinity between technology and the great political revolution of modern times.

either its social or its property relations, will be required
to achieve an economy of abundance. To Walker the machine represents the possibility of plenty shared by all.[27]

On this native political ground Walker develops his
most effective answer to Carlyle. Not that Carlyle had
denied the machine's power to improve the physical conditions of life: indeed, he had conceded that the nascent
industrial society probably would excel all others in the
"management of external things." But for him this was
not a damaging concession; he was convinced that the
material gains would be offset by "moral" losses, and that
the new system (which he would name "industrialism")
threatened to be the worst ever known with respect to
"pure moral nature," or "true dignity of soul and character." But, to repeat, Carlyle's language was not adequate
to his thought. In retrospect we can see that he was trying
to describe a condition of man's inner life, later to be
called alienation, which he regarded as characteristic of
the new, industrial environment. In any case, all of this
was lost upon Walker, who thought Carlyle's argument
typical of the pettifogging of pious reactionaries.

The distinction between the "external" and "internal"
consequences of mechanization, moreover, gave Walker
just the opening he needed. To the bright young lawyer,
it sounded like obfuscation or "mysticism," the weakest
spot in his adversary's position. And so, in a sense, it was.
To support his charge that science and technology were
producing a moral and spiritual decline, Carlyle (like
Schiller before him) had compared the quality of contemporary culture with that of ancient Greece. Walker
firmly rejects the comparison. True, he says, the Greeks
had achieved "high intellectual supremacy" without the
help of mechanism. But how had they done it? His reply

to his own question is simple, stunning, and tinged with just enough sans-culottish impertinence to give it bite:

> The answer . . . is ready. The Greeks themselves did not toil. Every reader of their history knows, that labor, physical labor, was stigmatized as a disgrace. Their wants were supplied by levying tribute upon all other nations, and keeping slaves to perform their drudgery at home. Hence their leisure. Force did for them, what machinery does for us. But what was the condition of the surrounding world? It is explained in a word. All other men had to labor for them; and as these derived no helps from Mechanism, manual labor consumed their whole lives. And hence their spiritual acquisitions have left no trace in history.

To the proposition that the new industrial order defiles the soul, Walker in effect is asking: whose soul, whose moral nature will be defiled? Would the souls of Greek slaves or English wage-laborers be threatened by labor-saving machines? All of which is a way of saying that men respond to change according to their social and economic perspective. Unlike Carlyle, the young American recognizes that mechanization will hardly seem a menace to those upon whom society confers little dignity of soul (or status) in the first place. To them the alternative is a life of drudgery or tedious, repetitive labor. On this point Walker is at his best. He addresses Carlyle in the manly voice of a humane, resolute, and thoroughly committed democrat. His political convictions in turn are supported by the scientific, humanistic faith of the Enlightenment. Like Jefferson, he assumes that knowledge can make us free, and not just some of us. It is all very well for Carlyle, or any European traditionalist, to compare the coming machine-based culture with ancient Athens, but let it be borne in mind, says Walker, "that we are speaking of society in the mass, and that our doctrine is, that men must be released from the bondage of perpetual bodily toil, before they can make great spiritual attainments."

To have freedom of mind, Walker is saying, it is first necessary to have freedom from want. It is to his credit, moreover, that he does not shy away from the criticism usually leveled against egalitarians on this point. He freely admits that the intellectual caliber of the mass, industrial culture, particularly at the higher levels, may well prove to be inferior to that of traditional cultures. But he is willing to take that risk. The kind of community he wants "may not produce a Newton, Milton, or Shakspeare," he says, "but it will have a mass of thought, reflection, study, and contemplation perpetually at work all over its surface, and producing all the fruits of mental activity." To the American lawyer, the machine is a token of the possibilities of democracy. It promises unbelievable abundance, hence a more harmonious and just way of life — including the life of the mind — than mankind ever has enjoyed.

5

Between 1786, when Thomas Jefferson described the new mill at Blackfriar's Bridge, and 1831, when Timothy Walker defended mechanization against Carlyle's attack, the image of the machine had become a major symbol of value. We call Jefferson's reference to the Boulton-Watt engine an "image" because it conveys little more than a sense perception. But in the speeches of Tench Coxe, the following year, the image is well on its way to becoming a new sort of "symbol." There it is made to carry a burden of implication, thought, and feeling far beyond that borne by a simple reference. Although it is not easy to draw a sharp line between image and symbol, there can be no doubt about the status of the machine in Timothy Walker's "Defence." There it is clearly, unmistakably, blatantly a cardinal symbol of value. It is not Walker's private symbol, but a property of the general culture. Within a few years it was to be as omnipresent as the image of the garden.

By 1844 the machine had captured the public imagination. The invention of the steamboat had been exciting, but it was nothing compared to the railroad. In the 1830's the locomotive, an iron horse or fire-Titan, is becoming a kind of national obsession. It is the embodiment of the age, an instrument of power, speed, noise, fire, iron, smoke — at once a testament to the will of man rising over natural obstacles, and, yet, confined by its iron rails to a predetermined path, it suggests a new sort of fate. The "industrial revolution incarnate" one economic historian has called it. Stories about railroad projects, railroad accidents, railroad profits, railroad speed fill the press; the fascinating subject is taken up in songs, political speeches, and magazine articles, both factual and fictional. In the leading magazines, writers elaborate upon the themes of Walker's essay. They adduce the power of machines (steam engines, factories, railroads, and, after 1844, the telegraph) as the conclusive sanction for faith in the unceasing progress of mankind. Associated with what seemed a world-wide surge of the poor and propertyless, with democratic egalitarianism, the machine is used to figure an unprecedented release of human energy in science, politics, and everyday life. Armed with this new power, mankind is now able, for the first time, to realize the dream of abundance. The entire corpus of intoxicated prose seems to rest on the simple but irresistible logic of first things first: all other hopes, for peace, equality, freedom, and happiness, are felt to rest upon technology. The fable of Prometheus is invoked on all sides. In his essay on "History" (1841), Emerson uses the example of the fire-stealer to suggest how "advancing man" unveils the authentic facts, such as the "invention of the mechanic arts," beneath the surface of ancient myth. "What a range of meanings and what perpetual pertinence," he says, "has the story of Prometheus!" [28]

But no summary in paraphrase will convey the subtle influence of industrialization upon mass consciousness. Not that this familiar and all-too-simple body of ideas requires further elucidation. Anyone can understand it, and, of course, that is just the point: it is the obviousness and simplicity of the machine as a symbol of progress that accounts for its astonishing power. With the possible exception of Henry Adams, no one grasped this fact so firmly as John Stuart Mill. In his brilliant comments (1840) on Tocqueville's *Democracy in America*, he argues that machine technology inculcates its message directly, imagistically, wordlessly. A locomotive is a perfect symbol because its meaning need not be attached to it by a poet; it is inherent in its physical attributes. To see a powerful, efficient machine in the landscape is to know the superiority of the present to the past. If the landscape happens to be wild or uncultivated, and if the observer is a man who knows what it means to live by physical labor, the effect will be even more dramatic and the meaning more obvious. "The mere visible fruits of scientific progress," says Mill, ". . . the mechanical improvements, the steam-engines, the railroads, carry the feeling of admiration for modern, and disrespect for ancient times, down even to the wholly uneducated classes." During the nineteenth century, therefore, no one needs to spell out the idea of progress to Americans. They can see it, hear it, and, in a manner of speaking, feel it as the idea of history most nearly analogous to the rising tempo of life.[29]

Much the same feeling surrounds the symbol of the machine when it is put into words. That is why it is not enough to sum up the "ideas" embodied in the symbol if we are to appreciate its impact upon the serious writers of the age. For its meaning is carried not so much by express ideas as by the evocative quality of the language, by attitude and tone. All of the writers of our first significant literary generation — that of Emerson and Hawthorne — knew this tone. It was the dominant tone of public rhetoric. They grew up with it; it was in their heads; and in one way or another they all responded to it. It forms a kind of undertone for the serious writing of the period, sometimes rising to the surface spontaneously, the writer momentarily sharing the prevailing ebullience, sometimes brought there by design for satiric or ironic purposes. In its purest form we hear the tone in Emerson's more exuberant flights; but it also turns up in Thoreau's witty parodies, in Melville's (Ahab's) bombast, in Hawthorne's satires on the age, and in Whitman's strutting gab and brag. To say this is not enough, however; one must hear the words, for their meaning is inseparable from the texture — the diction, cadence, imagery or, in a word, from the "language." Here, then, are some passages culled from the magazines of the period; they are offered not as a precise gauge of opinion, but rather as a sample of the rhetoric of progress surrounding the symbol of the machine.

1. *The machine and nature.* We live in an age marked by the "successful culture of physical science and . . . extraordinary triumphs in mechanism." Man has arrrived. "Indeed it would almost seem as though he were now but just entering on that dominion over the earth, which was assigned to him at the beginning. No longer, as once, does he stand trembling amid the forces of nature. . . ." To understand the American consciousness in this period the key image, as Tocqueville noted, is the "march" of the nation across the wilds, "draining swamps, turning the course of rivers, peopling solitudes, and subduing nature." Or, in the words of the writer quoted above:

> The wide air and deep waters, the tall mountains, the outstretched plains and the earth's deep caverns, are become parcel of his [man's] domain and yield freely of their treasures to his researches and toils. The terrible ocean . . . conveys . . . [him] submissively. . . . He has almost annihilated space and time. He yokes to his car fire and water, those unappeasable foes, and flying from place to place with the speed of thought carries with him, in one mass, commodities for supplying a province.

No stock phrase in the entire lexicon of progress appears more often than the "annihilation of space and time," borrowed from one of Pope's relatively obscure poems ("Ye Gods! annihilate but space and time, / And make two lovers happy."). The extravagance of this sentiment apparently is felt to match the sublimity of technological progress. "There appears to be something in the pursuit of mechanical invention," says one writer for the *Scientific American,* "which has a reaching up after our divine title, 'lords of the creation.' . . . It is truly a sublime sight to behold a machine performing nearly all the functions of a rational being. . . ." [30]

The entire relation between man and nature is being transformed. Discussing the moral and other indirect influences of the new railroads in 1832, one writer foresees an unprecedented harmony between art and nature, city and country. The American population of the future, he says, will "possess a large share of the knowledge, refinement, and polish of a city, united to the virtue and purity of the country." It is the new mechanized landscape itself which may be expected to induce this ideal state of mind. In explaining how the sight of a railroad will inspire future generations, this writer reveals the assumptions lying back of the progressive rhetoric, which also may be called the rhetoric of the technological sublime:

> Objects of exalted power and grandeur elevate the mind that seriously dwells on them, and impart to it greater compass and strength. Alpine scenery and an embattled ocean deepen contemplation, and give their own sublimity to the conceptions of beholders. The same will be true of our system of Rail-roads. Its vastness and magnificence will prove communicable, and add to the standard of the intellect of our country.[31]

Even the animals seem to recognize that a radical change is in store. Just back from an excursion on a new section of railroad, the editor of the *Cincinnati Enquirer* reports, in 1846, that he saw "herds of cattle, sheep, and horses, stand for a few seconds and gaze at the passing train, then turn and run for a few rods with all possible speed, stop and look again with eyes distended, and head and ears erect, seemingly so frightened at the tramp of the iron horse as to have lost the power of locomotion." People, he says, also were "dumfounded at the strange and unusual spectacle," and he often saw them rushing from their houses and "gaping with wonder and astonishment at the new, and to them grand and terrfic [sic] sight." Incredulity is a recurrent note. As if keyed to the subject, the rhetoric rises with man's power over nature.

> Steam is annihilating space. . . . Travelling is changed from an isolated pilgrimage to a kind of triumphal procession. . . . Caravans of voyagers are now winding as it were, on the wings of the wind, round the habitable globe. Here they glide over cultivated acres on rods of iron, and there they rise and fall on the bosom of the deep, leaving behind them a foaming wheel-track like the chariot-path of a sea-god. . . .

Written in 1844, this passage, with its image of the ship's wake as a "foaming wheel-track," anticipates the scene in *Moby-Dick* where Ahab, having asserted his dominion over the crew, identifies his will with the power of machines: he gazes astern and the *Pequod*'s wake seems to him the track of a railroad crossing the continent. His feeling for machines is not unlike that expressed by this writer for a businessman's magazine in 1840:

> We believe that the steam engine, upon land, is to be one of the most valuable agents of the present age, because it is swifter than the greyhound, and powerful as a thousand horses; because it has no passions and no motives; because it is guided by its directors; because it runs and never tires; because it may be applied to so many uses, and expanded to any strength.[32]

On every hand man is displaying titanic powers. The sight of a new bridge provokes one writer to ask, "What is there yet to be done upon the face of the earth, that cannot be effected by the powers of the human mind . . .?" The answer is that man ". . . is indeed, 'lord of creation'; and all nature, as though daily more sensible of the conquest, is progressively making less and less resistance to his dominion." [33]

2. *The machine and history.* The idea that history is a record of more or less continuous progress had become popular during the eighteenth century, but chiefly among the educated. Associated with achievements of Newtonian mechanics, the idea remained abstract and relatively inaccessible. But with rapid industrialization, the notion of progress became palpable; "improvements" were visible to everyone. During the nineteenth century, accordingly, the awe and reverence once reserved for the Deity and later bestowed upon the visible landscape is directed toward technology or, rather, the technological conquest of matter.

> The progress of human knowledge has accomplished within a century revolutions in the character and condition of the human race so beautiful and sublime as to excite in every observing mind feelings mingled with the deepest admiration and astonishment. No age has illustrated so strongly as the present the empire of mind over matter — and the ability of man to rise . . . above the obstacles with which nature has surrounded him. . . . It is a happy privilege we enjoy of living in an age, which for its inventions and discoveries, its improvement in intelligence and virtue, stands without a rival in the history of the world. . . . Look at our splendid steamboats. . . .

To look at a steamboat, in other words, is to *see* the sublime progress of the race. Variations on the theme are endless; only the slightest suggestion is needed to elevate a machine into a "type" of progress. Thus George Ripley in 1846:

> The age that is to witness a rail road between the Atlantic and Pacific, as a grand material type of the unity of nations, will also behold a social organization, productive of moral and spiritual results, whose sublime

and beneficent character will eclipse even the glory of those colossal achievements which send messengers of fire over the mountain tops, and connect ocean with ocean by iron and granite bands.[34]

But if the artifacts are new, the underlying assumptions remain those of the Enlightenment. Handbooks of intellectual history neglect the fact that the "romantic reaction" against scientific rationalism, the attitude of Poe, Hawthorne, or Melville, is directed not only against the dominant ideas of the previous age, but also against their omnipresence in contemporary culture. In the period between 1830 and 1860 popular discussions of technological progress assume that inventors are uncovering the ultimate structural principles of the universe. In 1850 a writer inspired by a new telescope says: "How wonderful the process by which the human brain, in its casket of bone, can alone establish such remote and transcendental truths." The overblown, exclamatory tone of so much of this writing arises from an intoxicated feeling of unlimited possibility. History has a meaning, a purpose, and a reachable goal: it is nothing less than man's acquisition of the absolute truth. Can we doubt, asks the same writer, that it is "the Divine plan, that man shall yet discover the whole scheme of the visible universe . . . ?" Of course, it may be said that in this case the language is not entirely inappropriate, since the subject is an astronomical discovery. But the hyperbolic rhetoric was not reserved for new penetrations of the heavens, as the following question, provoked by the appearance of an improved haymaker, so eloquently attests: "Are not our inventors absolutely ushering in the very dawn of the millennium?" [35]

3. *The machine and mind.* Describing the "Moral Influence of Steam" in 1846, a writer notes that it is "in the United States that the infant Hercules found a congenial

atmosphere, and imbibed that vigor which has since char-
acterised his labors and his triumphs." The main point of
the essay is that steam power, by relieving man from physical work, will upset the "moral economy of the world":

> . . . it is now universally employed as the great motive
> agent in machinery, triumphing over time and space,
> outstripping the winds in speed, annihilating every ob-
> stacle by sea or land, and almost defying the organic
> influences which regulate the surface of our globe. Nor
> is it only over matter that it exercises this control; for
> so wonderfully does it relieve the necessity of physical
> exertion, that it seems destined, in its future action and
> developments, to disturb the moral economy of the
> world, by opposing that great law of the universe, which
> makes labor the portion of man, and condemns him to
> earn his bread by the sweat of his brow. . . .[36]

Inventors, accordingly, are the intellectual heroes of the
age. Speaking of Eli Whitney's contribution, a visitor to
Yale in 1831 attacks those who feel that it "ought not to
be compared with what has been done by the intellectual
benefactors of mankind; the Miltons, the Shakspeares,
and the Newtons."

> But is it quite certain, that any thing short of the highest
> intellectual vigor, — the brightest genius, — is sufficient to
> invent one of these extraordinary machines? Place a
> common mind before an oration of Cicero, and a steam
> engine, and it will despair of rivalling the latter as much
> as the former. . . . And then, as to the effect on society,
> the machine, it is true, operates, in the first instance, on
> mere physical elements. . . . But do not all the arts of
> civilization follow in the train?

Later the rhetoric of progress becomes less diffident. One
recurrent assertion is that all the other arts rest upon the
mechanic arts. Another is that mechanism is, in itself, as

worthy as the fine arts: "inventions are the poetry of physical science, and inventors are the poets." To counter the sneers of literary traditionalists these writers develop ingenious turns. A favorite tactic is to borrow literary language, thereby producing such popular figures as "epic of machinery." "Who can tell of the dreamings — the wakeful nightly dreamings of inventors, their abstractions and enthusiastic reveries, to create some ballad or produce some epic in machinery."

> A writer on the steam-engine . . . declared that it required very much the same kind of genius and intellect to invent a new machine, as was necessary for the inspiration of a poem, and whether a man be a poet or inventor of machinery, is more the result of circumstances, or the age in which he chances to live, than in a difference of mental organization.[37]

A similar campaign is conducted on behalf of the machines themselves. Some people, one writer observes, regard the "mechanic arts as something undignified and degrading. . . ."

> Reader, if ever you have such feelings, go to Matteawan to be cured of them. When you enter the machine shop, you will understand why the Greeks, with their fine imagination, wedded Venus the beautiful to Vulcan of the hammer and the dusky brow. We always feel an unutterable pleasure in looking upon the operations of fine machinery. . . .

There is a defensive tone about much of this writing, in part a resistance to the exclusion of the "mechanic arts" from the general category of the "arts," which formerly had included all skills but in the early nineteenth century is increasingly reserved for the polite or "fine" arts. Raymond Williams has noted that the words "art" and "culture," as we use them, were in large measure formed in reaction against industrialization, and they still carry a strong negative bias. But not all literary intellectuals adopted the snobbish attitude. Emerson's essay on "Art" (1841) is written against it, and contributors to the *Dial,* in this case, Theodore Parker, often speak in this vein:

> The head saves the hands. It invents machines, which, doing the work of many hands, will at last set free a large portion of leisure time from slavery to the elements. The brute forces of nature lie waiting man's command, and ready to serve him.

Parker continues, offering what is probably the most popular conceit of the progressive vocabulary:

> At the voice of Genius, the river consents to turn his wheel, and weave and spin for the antipodes. The mine sends him iron Vassals, to toil in cold and heat. Fire and Water embrace at his bidding, and a new servant is born, which will fetch and carry at his command; will face down all the storms of the Atlantic; will forge anchors, and spin gossamer threads, and run errands up and down the continent with men and women on his back. This last child of Science, though yet a stripling and in leading strings, is already a stout giant. The Fable of Orpheus is a true story in our times.[38]

The idea that machine power is fulfilling an ancient mythic prophecy evokes some of the most exuberant writing. Under the improbable title "Statistics and Speculations Concerning the Pacific Railroad," we hear this excited voice: " '*By a horse shall Ilium perish,*' ran the prophecy in Old World times. 'By a horse America shall live,' saith the Oracle to the New World." The rhetoric often reaches its highest pitch when the machine is described hurtling in triumph across the landscape:

> And the Iron Horse, the earth-shaker, the fire-breather, which tramples down the hills, which outruns the laggard winds, which leaps over the rivers, which grinds the rocks to powder and breaks down the gates of the mountains, he too shall build an empire and an epic. Shall not solitudes and waste places cry for gladness at his coming?

Or, in an essay on the coal business of the United States, we are told that civilized society is indebted to fire "for the greatest portion of its superiority over savage life." To indicate the importance of fire the writer notes that "in the Grecian mythology . . . the daring theft of Prometheus" was so resented by the gods that the "punishment of its author was destined to be eternal, and terrible, in sublime horror, above all the retributive punishments of paganism." [39]

By the 1850's the celebrants of the machine take the offensive. Parity is not enough. A writer for the *Scientific American* regrets the "mass of thought and intelligence . . . expended by those who are termed the 'most highly educated,' upon subjects which have no practical bearing on the welfare of man or the advancement of the useful arts." What is striking here is that the complaint is directed not only against classical education, but against theoretical science as well. What good are long-winded studies of the age of the earth? No good at all! Academic interests of this kind explain the fact that "men from the workshop" (he mentions Watt, Fulton, Bell, and Stevenson) "have turned the world upside down by their inventions, while the sages of Oxford and Cambridge have but added some new theorems to the Principia." A sharp note of anti-intellectualism gets into the industrial celebration in the 1850's, and the Commissioner of Patents, Thomas Ewbank, gives the doctrine of utility an official stamp. Indeed, nothing sums up the metaphysic of industrialism so well as a statement of his, quoted in this instance under the heading "Civilization, Inventors, Invention and the Arts": " 'His works proclaim his preference for the useful to the merely imaginative, and in truth it is in such, that the truly beautiful or sublime is to be found. A steamer is a mightier epic than the Illiad, — and Whitney, Jacquard and Blanchard might laugh even Virgil, Milton and Tasso to scorn.' " [40]

4. *The machine and America.* There is a special affinity between the machine and the new Republic. In the first place, the raw landscape is an ideal setting for technological progress. In 1850 an American magazine reprints an English writer's explanation for the greater success of technology, especially the telegraph, in the New World. Only half a century before, he says, "wild beasts, and still wilder Indians, wandered over the lands now traversed in perfect security by these frail wires. . . ." In some not wholly explicable way the backwardness of the country gives the progressive impulse an electric charge; hence the "transition from a wild and barbarous condition to that of the most elaborate civilization . . . [is not] gradual, but instantaneous." In America progress is a kind of explosion. Civilization "has at one bound leaped into life, surrounded with every appliance . . . which the existing knowledge of man has devised for ministering to his wants and his enjoyment." Never has there been anything like the violent coming together of advanced art and savage nature. Nor is the singularity of this clash lost upon Americans. In the words of George Perkins Marsh, speaking to the Rutland County Agricultural Society in 1847,

America is the "first example of the struggle between civilized man and barbarous uncultivated nature." Elsewhere
the earth has been subdued slowly, but here for the first
time "the full energies of advanced European civilization,
stimulated by its artificial wants and guided by its accumulated intelligence, were brought to bear at once on a desert
continent. . . ." [41]

But there is an equally important political reason for
the astonishing success of mechanism in the New World.
After all, well-trained, ingenious minds had confronted
barbarous conditions before. Consider the ancients, the
exemplary Greeks and Romans. Although they had had
brilliant, inventive minds, their inventiveness was "wrongly
directed." They were not interested in the useful arts,
chiefly because of what a twentieth-century social scientist
would call their "value system," but what this writer in
1847 refers to as a "philosophy . . . repugnant to any
invention which had for its object the benefit of the mass
of mankind."

> This very element in ancient polity and ancient phi-
> losophy, was the very cause which prevented in the
> working classes of those days that developement [sic] of
> mechanical genius, which by the construction of ma-
> chines . . . might have elevated both Greece and Rome
> to that pinnacle of greatness on which some . . . mod-
> ern nations now stand. . . . Every effort of genius is
> prostituted, unless directed for the purpose of benefitting
> the human family.

To account for the progress of American technology it is
not enough to talk about geography, or even the combined
effect of the virgin land and Yankee inventive skill. One
must also recognize the incentives which call forth that
skill. They are provided by a democracy which invites
every man to enhance his own comfort and status. To the

citizen of a democracy inventions are vehicles for the pursuit of happiness.[42]

The result is that Americans have seized upon the machine as their birthright. Granted that the best of the fine arts — statuary, painting, and architecture — is still to be found in the Old World, "there is one agent which we can call peculiarly our own, and in the application of which, the nation is destined to excel." What is more, the agent has appeared at a providential moment, just when our manifest destiny requires it.

> Just as we are prepared to go forward in building the frame of our national enterprise, a new power presents itself! The spirit of the republic grasps it, . . . hails the agency of steam as the benefactor of man, and the power which stamps the character of the present age.

The new power, this benefactor of man, is not to be selfishly guarded as an American possession. (The European origins of industrial technology often are neglected.) Even now, in 1847, the inventive genius of the Republic is spreading its blessings in Europe. In Russia 30,000 men are building railroads under American supervision. Looking forward to the completion of this project, a writer extends the grand conceit of the progressive rhetoric, the image of a Promethean fire-machine sweeping across the continent and even beyond national frontiers, back to the Old World. He envisages the moment when, "across the Steppes of the Volga and through the Passes of the Ural Mountains, will yet roll the swift American locomotive, pealing notes of nobler victories than those of the reddest warfare—the triumphs of American mechanical genius."

> Who knows now what great and good influence in the cause of Freedom and Reform is exercised by the

mingling of our mechanics with the peasantry of the
Russian empire. — Who knows but in a few years the
now Russian serf, may stand a freeman at his own cot-
tage door, and as he beholds the locomotive fleeting past,
will take off his cap . . . and bless God that the Mechanics of Washington's land were permitted to scatter
the seeds of social freedom in benighted Russia.[43]

This little fantasy, which concludes our sample of progressive rhetoric, is a projection of feelings that pervade
the whole grandiloquent litany. By now the image of the
American machine has become a transcendent symbol: a
physical object invested with political and metaphysical
ideality. It rolls across Europe and Asia, liberating the
oppressed people of the Old World — a signal, in fact, for
the salvation of mankind. It fulfills, as Walt Whitman
later would put it, the old yearning for a "Passage to
India." Above all, the rhetoric conveys that sense of unlimited possibility which seizes the American imagination at
this remarkable juncture. Before coming down on the
meretriciousness of the language, we need to remind ourselves how remarkable, in truth, the circumstances were.
Consider how the spectacle of the machine in a virgin
land must have struck the mind. Like nothing ever seen
under the sun, it appears when needed most: when the
great west finally is open to massive settlement, when
democracy is triumphant and gold is discovered in California, here — as if by design — comes a new power commensurate with the golden opportunity of all history. Is
it any wonder that the prospect arouses awe and reverence?
Back of the stock epithets and the pious, oracular tone,
there are emotions which cannot be dismissed as mere
hokum: a plausible incredulity, wonder, elation, and
pride; a generous, humane delight at the promise of so

much energy so soon to be released. But this is not to deny the intellectual hollowness of the rhetoric. The stock response to the panorama of progress, as Mill observed, by-passes ideas; it is essentially a buoyant feeling generated without words or thought. The same may be said of the rhetoric. It rises like froth on a tide of exuberant self-regard, sweeping over all misgivings, problems, and contradictions.

And there were misgivings. Quite apart from any overt criticism of the new power, it is possible to detect tremors of doubt within the rhetoric of praise. Often writers use imagery which belies their arguments. In 1840 James H. Lanman made what seems on first inspection to be a wholly affirmative survey of the "Railroads of the United States." Adopting the booster tone of Hunt's *Merchants' Magazine*, he praises the new machines as "the triumphs of our own age, the laurels of mechanical philosophy, of untrammeled mind, and a liberal commerce!" It is clear, he says of the railroads, "that all patriotic and right-minded men have concurred in the propriety of their construction." Nevertheless, he manages to convey something less than full confidence in the benign influence of machines. They are, he says, "iron monsters," "dragons of mightier power, with iron muscles that never tire, breathing smoke and flame through their blackened lungs, feeding upon wood and water, outrunning the race horse. . . ." At one point he describes a train "leaping forward like some black monster, upon its iron path, by the light of the fire and smoke which it vomits forth." All of these images, which associate technology with the destructive and repulsive, are in marked contrast to Lanman's manifest theme. They communicate a sense of anxiety and dread. It may be significant that he manages to make technology seem most loathsome in describing the effect of a "steam screw" upon the landscape. It is an instrument, he says, which "should tear up by the roots the present monarchs of the forest, and open the ample bosom of the soil to the genial beams of the fertilizing sun." [44]

But aside from such covert expressions of ambivalence, everyone knows that the great majority of Americans welcomed the new technology. As Perry Miller said, welcomed is too weak a verb: they grasped and panted and cried for it. Again and again, foreign travelers in this period testify to the nation's obsessive interest in power machinery. The typical American, says Michael Chevalier, "has a perfect passion for railroads; he loves them . . . as a lover loves his mistress." In the words of another Frenchman, Guillaume Poussin, "the railroad, animated by its powerful locomotive, appears to be the personification of the American. The one seems to hear and understand the other — to have been made for the other — to be indispensable to the other." Perhaps the most interesting comment, like a report of an informal projective test, comes from Frederika Bremer. When the boys in a schoolroom she is visiting are told to amuse themselves drawing on their slates, the Swedish novelist notices that most of them draw "smoking steam-engines, or steamboats, all in movement." As if endorsing Carlyle's notion that mechanism is taking possession of mind, Miss Bremer concludes that "interest in locomotive machinery has a profound connection with life in this country." So profound indeed was the connection that Americans had little difficulty in reconciling their passion for machine power with the immensely popular Jeffersonian ideal of rural peace, simplicity, and contentment. As an example of the way that reconciliation was accomplished, let us consider an episode in the life of the nation's foremost political rhetorician.[45]

6

Because he owned a farm nearby, Daniel Webster happened to be in Grafton, New Hampshire, on August 28, 1847, the day the Northern Railroad was opened. A large crowd from neighboring farms and villages had gathered for the ceremonies. When the great orator's presence became known, he was called upon — spontaneously and with enthusiasm, we are told — and Mr. Webster "readily complied." Three months later, on November 17, another stretch of the railroad was opened in Lebanon. To celebrate this event a more formal entertainment, including a "collation," was prepared. Several distinguished guests were brought by rail all the way from Boston. When the official train reached South Franklin, it stopped to "take in" Mr. Webster. Later that day the assembly once again demanded a speech. "Thus called upon to speak," he said to the crowd, "I cannot disregard the summons." A master of language, keen politician, his ear nicely tuned to the prevailing mood, Webster knew exactly what was wanted. Taken together, these ceremonial speeches exhibit one way of neutralizing the conflict between the machine and the rural ideal.[46]

Webster's major theme, needless to say, is the progress exemplified by the railroad. At first he dwells upon the changes in the countryside since his youth, when there were no roads between the river valleys. "At that day," he says, "steam, as a motive power, acting on water and land, was thought of by nobody. . . ." Then came the remarkable series of internal improvements, first roads, then turnpikes and canals, and now, vastly greater, "the invention which distinguishes this age." The railroad "towers above all other inventions of this or the preceding age," he says, "as the Cardigan Mountain now before us lifts itself above the little hillocks at its base. Fellow-citizens, can we without wonder consider where we are, and what has brought us here?"

In answer to his own question, Webster brings on the familiar homage to progress. The railroad breaks down regional barriers. Those who came from Boston might have brought along "fish taken out of the sea at sunrise." Imagine, says Webster, eating as good a fish dinner in the mountains of New Hampshire as on the beach at Nahant! The new inventions hold the promise of national unity and, even more exciting, social equality. Nothing could be as important to the "great mass of the community" as this innovation "calculated . . . to equalize the condition of men." It is a mode of conveyance available to rich and poor alike, and he is pleased to report that the people regard it as their own. Upon asking one of his tenants ("my farmer") the meaning of a "line of shingles" across his fields, the man had replied, " 'It is the line of our railroad.' Our railroad!" Webster exclaims, "That is the way the people talked about it. . . . It is the spirit and influence of free labor, it is the indomitable industry of a free people, that has done all this."

But Webster is not satisfied to rehearse these stock themes; he also feels a responsibility to dispose of certain "idle prejudices" against railroads. Some he dismisses quickly. In one sentence he brushes aside the charge that the new companies are undemocratic, monopolistic, closed corporations. What else could they be? — "the track of a railway cannot be a road upon which every man may drive his own carriage." As for the danger of infringing on the rights of private property, that is easily avoided by ensuring ample compensation to landowners. Such technical problems apparently do not concern him. What does interest him, however, is a less easily defined way in which the railroads "interrupt or annoy" people:

> When the directors of the road resolved to lay it out upon the river (as I must say they were very wise in doing), they showed themselves a little too loving to me, coming so near my farm-house, that the thunder of their engines and the screams of their steam-whistles, to say nothing of other inconveniences, not a little disturbed the peace and repose of its occupants.

As the diction plainly indicates, Webster is aware of certain far-reaching if unspoken implications of the event. He heightens the emotional charge by employing the stock device of the monstrous machine with its "thunder" and "screams," and by setting it against the conventionally pastoral "peace and repose" of the farm. Then, as if the train roaring past his rural retreat were not inconvenience enough, he tells how the landscape is being desecrated. "There is, beside," he says, "an awkward and ugly embankment thrown up across my meadows. It injures the look of the fields." For a moment at least the audience may well have been puzzled. What is he saying? Is he really annoyed? Does this monster threaten the American way of life? Webster continues:

> It injures the look of the fields. But I have observed, fellow-citizens, that railroad directors and railroad projectors are no enthusiastic lovers of landscape beauty; a handsome field or lawn, beautiful copses, and all the

gorgeousness of forest scenery, pass for little in their
eyes. Their business is to cut and slash, to level or deface
a finely rounded field, and fill up beautifully winding
valleys. They are quite utilitarian in their creed and in
their practice. Their business is to make a good road.
They look upon a well-constructed embankment as an
agreeable work of art; they behold with delight a long,
deep cut through hard pan and rock, such as we have
just passed; and if they can find a fair reason to run a
tunnel under a deep mountain, they are half in raptures.
To be serious, Gentlemen . . .

The last, transitional phrase is the key to Webster's
strategy. Without it, in fact, we might have some difficulty
imagining his tone of voice. But it announces that the
offended sensibility of lovers of landscape, for whom he
had pretended to speak, is not to be taken seriously.
Chances are that the note of injury in this plaint was
delivered as mock injury. In any case, he is saying that
it is foolish to deplore the cutting and slashing. From an
eminence high above the offended and the offenders,
Webster speaks as one who sees all around the situation;
he sees the ugly scar cut into the green hills of New Hampshire, and he sees the forces represented by the railroad,
and he knows beyond any possible doubt that to those
forces Americans must and will pay homage. Within view,
as he speaks, there is a locomotive embodying the dominant impulse of the society: "a zealous determination to
improve and profit by labor. . . ." He admits that in their
zeal Americans may seem to neglect impulses other than
those resulting in profit or improvement. "New Hampshire, it is true," he says, "is no classic ground. She has no
Virgil and no Eclogues." But the common-sense doctrine
of first things first tacitly controls the entire discourse.

Pastoral poetry and the beauty of the landscape belong in one category, along with the peace and repose of rural life, but they are not to interfere with *serious* enterprises, with the activity of railroad promoters, men in touch with reality, whose "business is to make a good road." When Webster says, "To be serious, Gentlemen," it is with the serene assurance that his audience shares his definition of what is serious. "To be serious, Gentlemen," he explains, "I must say I admire the skill, the enterprise, and that rather bold defiance of expense . . . [of] the directors of this road. . . ."

With impressive oratorical craft, Webster has made an example of himself in order to display the proper response to the "inconveniences" attendant upon industrial progress. His trick is *reduction* in the technical literary sense of giving in to a feeling or idea in order, eventually, to take it back. By bringing it out into the open, he helps assuage that inchoate, hovering anxiety about change which often expresses itself obliquely, in moral or aesthetic language, and in a revulsion against the machine as a physical object. Actually, this feeling is grounded in the older, pastoral conception of American society associated with Jefferson's noble husbandman. By treating it as trivial, effete, and literary, Webster reduces the psychic dissonance generated by industrialization. To be truly serious, he says, is to put such feelings aside and rejoice in the changes wrought by the machine. Each speech, accordingly, ends on a note of high praise. At Grafton, in August, he urges that the policy which led to the building of the road be pursued "till internal improvement in some really and intrinsically useful form shall reach every glen and every mountain-side of the State." At Lebanon, in November, a more formal occasion for which he is better prepared, Webster ends with a Ciceronian tribute to the mechanized sublime:

> It is an extraordinary era in which we live. It is altogether new. The world has seen nothing like it before. I will not pretend, no one can pretend, to discern the end; but every body knows that the age is remarkable for scientific research into the heavens, the earth, and what is beneath the earth; and perhaps more remarkable still for the application of this scientific research to the pursuits of life. The ancients saw nothing like it. The moderns have seen nothing like it till the present generation. . . . We see the ocean navigated and the solid land traversed by steam power, and intelligence communicated by electricity. Truly this is almost a miraculous era. What is before us no one can say, what is upon us no one can hardly realize. The progress of the age has almost outstripped human belief; the future is known only to Omniscience.

Now, in the end, the orotund language overwhelms any lingering trace of uncertainty. The stately cadence of the peroration, a cliché the audience surely expects, lends the Senator's argument an aura of priestly assurance. The occasion has a certain ritual aspect to begin with, but now the rhetoric of the technological sublime heightens it, bringing in a sense of cosmic harmony, suggesting an obscure kinship between Webster, the spirit of the Republic, the machine power, and the progressive forces of history. Everything, it says, is working out according to a divine plan. Now the disturbing images of change, the screaming monster and the defacer of landscape, seem embarrassingly squeamish, effeminate, and trivial. The noise and smoke, the discomfort and visual ugliness, even the loss of peace and repose — these things, the rich voice proclaims, are of little consequence to true Americans.

That the Senator knew his countrymen is beyond question. Consider the increase in the nation's total railroad mileage before 1860: in 1830, 73 miles had been laid, in 1840, 3,328 miles, in 1850, 8,879 miles, and in 1860, 30,636 miles. But the contradiction between the feverish activity represented here and the pastoral ideal did not go undetected. Indeed, we shall be in a better position to appreciate the representative character of Webster's attitude if we set it beside its opposite. Consider, for example, a description of industrialization in Vermont printed in the Fourierist journal, the *Harbinger*, in 1847, the year of Webster's tributes to the Northern Railroad. The writer, who uses a pseudonym, almost certainly is one John Orvis, a recent resident of Brook Farm and a socialist lecturer. Just returned from a trip through his native state, Orvis begins with an apostrophe to the beauty of the scenery (the "luxuriant growth of grass upon the meadows . . . the freshest green upon the hill-side pastures"), and, in fact, he says, there is "no describing the picturesque harmony of the landscapes," which he thereupon describes:

> The countenance of these hills is of inimitable softness, whilst the numberless flocks and herds that animate them, prove how wisely use is married with beauty. It is sublime, to stand in the mighty bowling alley which these mountains form . . .

The terrain, in short, figures an ideal blending of art and nature; it is a version of the pastoral middle landscape.[47]

But in the last fifteen years, alas, this Virgilian countryside has been the scene of a technological and social revolution. First came the textile mills. They attracted cheap Canadian labor, and now even the daughters of native farmers "throng the factories . . . where they are al-

lowed but ten minutes to eat their dinners, and forced to
sleep in brick pens. . . ." What follows is a Dickensian
picture of calamitous industrialization — the girls "without home, friends or counseling, wearing life to decay,
and weaving themselves shrouds whilst earning a gown"
— but the worst is yet to come. Until now lack of transportation has prevented the use of the state's magnificent
factory sites; but that impediment soon is to be removed.
The great topic of excitement everywhere is railroad building. Within ten years, Orvis predicts, the "wild picturesque
waterfalls . . . will be deformed by the ugly presence of
mills, and their voices, that now sing to their mountain
dance, will then groan at the slavish wheel." The transformation of visible nature prefigures a social transformation; soon, he says, the "beautiful pastoral life of the inhabitants will give place to oppressive factory life — quiet,
rural pursuits will be absorbed in the din, conflict and
degradation of manufacturing and mechanical business.
. . ." Dire political changes will follow, the "golden
equality which now exists, will precipitate into rigorous
forms of caste, of capitalist and laborer." We may pause
a moment to notice how the suggestion of a golden age
helps to fix egalitarian social principles in a pre-industrial
landscape. The point is, says Orvis, that Vermont soon will
cease to be the "Evergreen State," and a "false society will
have blasted its beauty and dried up the blood of its
vigor and prosperity." (The blasting which destroys the
beauty of the state is the blasting which clears a path for
the new railroad.) Mechanization, both literally and metaphorically, means disharmony. It separates the people from
the lovely green landscape which has, or ought to have,
a primary place in their thought and feeling. Between
man and nature it threatens to impose an ugly, depressing, and inhumane community.

What needs emphasis here is that Orvis is insisting upon precisely the metaphor Webster casually dismisses; he uses the image of the machine in the native landscape to figure a dangerous contradiction of social value and purpose. But Webster, knowing his audience, brings in the technological sublime to neutralize the dissonance generated by industrialization. The rhetoric forms an emotional bond between the orator and the public. It puts him in touch with the mass surge toward comfort, status, wealth, and power that rules the society. If Orvis is correct when he argues the patent inconsistency between industrialization and pastoral ideals, Webster is correct when he says, in effect, that it doesn't matter. Orvis has got hold of an intellectual, logical, and literary truth, but Webster understands the practical political truth — the facts of power.

The contrast, a case example in the politics of American culture, roughly indicates the relative force exerted by the opposed symbols of the machine upon the native consciousness. It is fitting that the apostrophe to the new power should be offered by a national hero, closely associated with the nation's dominant political, industrial, and financial groups, and speaking under the auspices of a corporation: Webster represents the organized community. It is fitting, too, that the essay that calls into question the assumptions governing capitalist industrialism should be written by a little-known reformer under an assumed name in an ephemeral, obscure organ of the Brook Farm Associationists. In the public view he appears as one of those queer intellectuals who flaunt their disaffection with gestures of withdrawal and, indeed, by quite literally withdrawing from ordinary social life to rustic utopian communities, pastoral oases. (An ". . . illusion, a masquerade, a pastoral, a counterfeit Arcadia," says Hawthorne's narrator, Coverdale, of the fictive Brook Farm in

The Blithedale Romance.) If Webster stands forth as the type of solid Yankee gentility, property, eloquence, and position, Orvis brings to mind a small cult of literary dreamers beyond the fringe.[48]

The invidious comparison is further heightened by what might be called the public personality of the two ideologies, or world views, represented here. In the idiom of the 1840's, Orvis's attitude inevitably will be taken for another of the wild, "Germanic" theories to which the effete and the literary are said to be vulnerable. The epithet is calculated to evoke memories of Goethe and other vaguely disreputable poets with curious manners and an unrealistic, freewheeling, metaphysical turn of mind. It means, in a word, romantic *Weltschmerz*, a state of feeling thought to be basically subversive yet in most cases, like "beat" rebelliousness today, adolescent and harmless. To respectable, middle-class Americans who admire Daniel Webster, the viewpoint adopted by Orvis seems as mystical and un-American as Carlyle's "Signs of the Times" had seemed to Timothy Walker. Not that the difference between the two views is in any important sense an international difference; in England the division between the general culture and the intellectual, or literary, culture conforms to much the same pattern. In both countries the critical animus of men like Orvis or Carlyle was dismissed as the product of alien, European influences. But the case was stronger in America. Having been carried across the Atlantic on a wave of avant-garde, romantic thought, the new attitude seemed even more exotic in Vermont than in the Lake District. Here, after all, was a young Republic whose ideological underpinnings had been firmly set in the Enlightenment. Not only did the Constitution incorporate many of the basic assumptions of the "mechanists," but the nation's favorite philosopher remained, well into the nineteenth century, "the Great Mr. Locke." All of which helps to explain the superb confidence of Daniel Webster, patriot and native Demosthenes, who keeps alive the neo-classic image of the American statesman.[49]

Given the political and ideological climate, it is not surprising that the pastoral ideal is invoked against industrialization chiefly by those who, like Orvis, are radically disaffected.* Aside from apologists for Southern slavery, these dissidents belong to small, ineffectual groups — socialist, transcendentalist, literary, religious — far from the centers of influence and power. Outside the South the pastoral ideal has little or no practical value as a political weapon against industrialism. Even the Jacksonians, who have deep misgivings about what is happening to the nation, adopt an attitude much closer to Webster's. To be sure, they attack the Bank as a monstrous piece of financial machinery (Jefferson had referred to the Hamiltonian system as a machine), and they attempt in various ways to stay the centripetal forces at work in the economy.

* Yet the overt negative response to industrialization is not nearly as rare, in the written record, as one might expect. That is because it has a special appeal for the more literate and literary, hence it appears in print with a frequency out of all proportion to its apparent popularity with the public. One might easily cull an impressive exhibit of passages displaying hostility to the new technology (a kind of composite image of the menacing machine to set beside the sublime machine) from the periodicals of the day. But a merely quantitative sample of periodical literature would be misleading. By and large the negative response appears in newspapers like the *Harbinger*, organs of small organized minorities. On the other hand, the rhetoric of the technological sublime appears chiefly in such respectable magazines as Hunt's *Merchants' Magazine*, *Littell's Living Age*, *Niles' Register*, the *North American Review*, and the *Scientific American*, journals which on the whole represent the views of the governmental, business, and professional elites.

But in politics, the swelling, surging demand for everything that technology promises is irresistible. The result, as John William Ward and Marvin Meyers demonstrate, is that the Jacksonian "persuasion" embraces a typically nostalgic, or, as we say, ambivalent, look-both-ways kind of native progressivism. For all their misgivings, the Jacksonians are no more inclined than Webster to insist upon a root contradiction between industrial progress and the older, chaste image of a green Republic. On the whole they share the prevailing assumption that machine technology (and all that it represents) belongs, or can be made to belong, in the middle landscape.[50]

The classic pictorial statement of this major theme is provided by George Inness. In 1854, seven years after Webster's New Hampshire speeches, the Lackawanna Railroad Company commissioned Inness to paint the scene of its operations. At first, apparently, the assignment repelled him. Hitherto a painter of pleasant, but on the whole conventionally romantic, landscapes, Inness was put off by the notion of painting anything as devoid of visual charm as a repair shop, a roundhouse, or a smoking locomotive. He did not see how such objects could be assimilated to his habitual Virgilian mode. But as it happens the difficulty proved to be stimulating, and the result, *The Lackawanna Valley* (1855; plate 2), is generally included among his best works.* It is a striking representation of the idea that machine technology is a proper part of the landscape. Like Thoreau's extended metaphor at the end

---

* One scholar, Wolfgang Börn, believes that Inness's effort to impose order upon this novel subject matter produced a significant technical innovation. In Börn's view, the striking luminosity of the work, which he thinks at least ten years in advance of continental palettes, was devised specifically as a means of making the technological artifacts seem of a piece with the rest of the landscape.

PLATE 2

"The
Lackawanna
Valley"
by
George Inness

PLATE 3

"American Landscape"
by Charles Sheeler

of *Walden,* the springtime thawing of the sand in the rail-road bank, Inness's painting seems to say that "there is nothing inorganic." Instead of causing disharmony, the train is a unifying device. The hills in the background and the trees of the middle distance gently envelop the industrial buildings and artifacts. No sharp lines set off the man-made from the natural terrain. Nor is the Lacka-wanna's smoke unpleasant. The cottony puffs that rise from the engine and the roundhouse are merely duplicates of a puff that rises from behind the church — an ingenious touch! Instead of cutting the space into sharp, rectilinear segments as railroad tracks often do, the right-of-way curves gracefully across the center of the canvas where it divides in two, forming the delicately touching ovals that dominate the middle plane. It is noteworthy, too, that the animals in the pasture continue to graze or rest peacefully as the tidy, diminutive train approaches. Still, this is not a lament for Goldsmith's cherished land; the stumps indi-cate that the pasture has just been hewn out of a wilder-ness. But, of course, it is the solitary figure reclining beneath the dominant vertical, the tree in the foreground, who finally establishes the quiet, relaxed mood. He holds no crook, but he contemplates the sight in the serene pos-ture of the good shepherd looking out across Arcadia.[51]

In 1855, the year that Inness completed his painting, Thomas Buchanan Read, a Connecticut poetaster, pub-lished his long poem, *The New Pastoral.* He begins with an invocation to the land, and especially to

> . . . that pastoral phase,
> Where man is native to his sphere, which shows
> The simple light of nature, fresh from God! —
> That middle life, between the hut and palace,
> 'Twixt squalid ignorance and splendid vice; —

> Above, by many roods of moral moves,
> The Indian's want, and happily below —
> If the superior may be called below —
> The purple and fine linen. . . .

He sings of "iron men who build the golden future," men with "heroic wills, . . . / . . . to which / The Wilderness, the rank and noxious swamps, / . . . bow and bear / The burthen of the harvest." Like Inness, Read is invoking the image of the middle landscape, a version of the pastoral ideal which continues to engage the American imagination long after the "take-off" into the industrial age.[52]

Of American writers with authentic gifts, Emerson and Whitman pay the most direct, wholehearted tribute to this industrialized version of the pastoral ideal. It is fitting to consider Whitman here because he comes closest to transmuting the rhetoric of the technological sublime into poetry. Although the career of his mythic American, the "I" of "Song of Myself," begins with the contemplation of a simplified, stripped-down natural landscape ("I lean and loafe at my ease observing a spear of summer grass"), he has no difficulty assimilating the forces represented by the machine. In "Crossing Brooklyn Ferry" there is a moment of hesitation, just a touch of doubt, as he notes the black and red lights cast by the foundry chimneys over the houses and streets. Later in the poem the industrial blackness is linked to an inward state, the "dark patches" of evil he discovers within himself. But the conflict is easily resolved, and in the final stanza the speaker confidently returns to the theme in the imperative voice:

> Burn high your fires, foundry chimneys! cast black
>   shadows at nightfall! cast red and yellow light
>   over the tops of the houses!

Here, as elsewhere, the American scene induces an exuberance in Whitman's hero that rises above all possible doubts.

Although it comes after the Civil War, Whitman's "Passage to India" (1868) probably is the purest, most poetic use of the progressive idiom in our literature. It begins:

> Singing my days,
> Singing the great achievements of the present,
> Singing the strong light works of engineers,
> Our modern wonders, (the antique ponderous Seven
>   outvied,)
> In the Old World the east the Suez Canal,
> The New by its mighty railroad spann'd,
> The seas inlaid with eloquent gentle wires;
> Yet first to sound, and ever sound, the cry with thee
>   O soul,
> The Past! the Past! the Past!

Against the forward thrust of mechanized change, the mild "yet" counters a yearning for the past. But this feeling is soon overwhelmed by the sense of buoyant power that arises from the sight of the machine's motion across the landscape:

> I see over my own continent the Pacific railroad
>   surmounting every barrier,
> I see continual trains of cars winding along the
>   Platte carrying freight and passengers,
> I hear the locomotives rushing and roaring, and the
>   shrill steam-whistle,
> I hear the echoes reverberate through the grandest
>   scenery in the world. . . .

And so on. Like the writers for the popular magazines, the poet foresees all the continents spanned, connected by one network, "welded together," and the globe one

> . . . vast Rondure, swimming in space,
> Cover'd all over with visible power and beauty. . . .

As for the "yet," the speaker's sense of separation from the past, his fear that he will forget his origins (most forcibly expressed as a cry out of the "dark unfathom'd retrospect"), it is dissolved by the familiar, all-answering logic of first things first.

> After the seas are all cross'd, (as they seem already
>   cross'd,)
> After the great captains and engineers have accomplish'd
>   their work,
> After the noble inventors, after the scientists,
>   the chemist, the geologist, ethnologist,
> Finally shall come the poet worthy that name,
> The true son of God shall come singing his songs.

In "Passage to India" the machine power is a precursor of a higher, spiritual power. *After* it has done its work the divine bard will arrive to announce that "Nature and Man shall be disjoin'd and diffused no more." His songs will confirm the speaker's prevision of history as an upward spiral, a movement that dispels all doubt, carrying mankind back, full circle, to the simple vitality of "primal thought," above the "gardens of Asia" where history began:

> . . . the voyage of his mind's return,
> To reason's early paradise,
> Back, back to wisdom's birth, to innocent intuitions,
> Again with fair creation.

It should be said that "Passage to India" lacks the sharp particularity which redeems the bombast in much of Whitman's early work. Perhaps the hinted conflict between technological progress and the garden ideal, impossible to resolve on the plane of ordinary experience, lends a particular impetus to this declamatory manner.

In "Passage to India" Whitman is expressing attitudes that dominated the general culture at the time. So far from causing serious concern, the new power often is interpreted as a means of realizing the classical, eighteenth-century aims of the Republic. According to a writer for a business magazine, factory labor is not "pursued here as in England." There it is "a continuous business for life," but here the young men and women enter factories "not as a main object of pursuit, but as a stepping-stone to a future settlement. . . ." After they have earned a small sum they are likely to "emigrate to the broad and rich fields of the west, where the soil, like a kind mother, opens its arms to receive them, and where they settle down permanent freeholders, perhaps the future legislators of the country." It is technology, indeed, that is creating the new garden of the world. "The great Mississippi valley," says a writer in *DeBow's Review* in 1850, "may emphatically be said to be the creation of the *steam engine,* for without its magic power . . . what centuries must have elapsed before the progress of arts and of enterprise could have swept away the traces of savage life." The railroad is the chosen vehicle for bringing America into its own as a pastoral utopia. That it also means planting Kansas City where the garden was supposed to be does not often occur to the popular rhetoricians.[53]

No one captures the dominant mood so well as a writer for the *Scientific American* in 1849. Addressing himself to the question, "What Is the Golden Age," he begins by attacking those whose "affections are all with the past." Then, in the familiar idiom, he reviews the progress of the arts and sciences. They are doing more than has ever been done before to "render mankind virtuous and happy." But surprisingly enough, his conclusion is not that the golden age lies in the future. Rather, he says, in a delightfully bland and circumstantial tone, "so far as we can judge, in looking back upon the experience of our life in the world, our opinion is favorable to the *now* being the *golden age*." [54]

The pastoral idea of America had, of course, lent itself to this illusion from the beginning. In the eighteenth century it had embraced a strangely ambiguous idea of history. It then had provided a clear sanction for the conquest of the wilderness, for improving upon raw nature and for economic and technological development — up to a point. The objective, in theory at least, was a society of the middle landscape, a rural nation exhibiting a happy balance of art and nature. But no one, not even Jefferson, had been able to identify the point of arrest, the critical moment when the tilt might be expected and progress cease to be progress. As time went on, accordingly, the idea became more vague, a rhetorical formula rather than a conception of society, and an increasingly transparent and jejune expression of the national preference for having it both ways. In this sentimental guise the pastoral ideal remained of service long after the machine's appearance in the landscape. It enabled the nation to continue defining its purpose as the pursuit of rural happiness while devoting itself to productivity, wealth, and power. It remained for our serious writers to discover the meaning inherent in the contradiction.

Then the little locomotive shrieked and began to move: a
rapid churning of exhaust, a lethargic deliberate clashing of
slack couplings travelling backward along the train, the exhaust changing to the deep slow clapping bites of power as
the caboose too began to move and from the cupola he
watched the train's head complete the first and only curve
in the entire line's length and vanish into the wilderness,
dragging its length of train behind it so that it resembled a
small dingy harmless snake vanishing into weeds, drawing him
with it too until soon it ran once more at its maximum clat-
tering speed between the twin walls of unaxed wilderness as
of old. It had been harmless once. . . .

It had been harmless then. . . . But it was different now.
It was the same train, engine cars and caboose . . . running
with its same illusion of frantic rapidity between the same
twin walls of impenetrable and impervious woods . . . yet
this time it was as though the train . . . had brought with
it into the doomed wilderness even before the actual axe the
shadow and portent of the new mill not even finished yet
and the rails and ties which were not even laid; and he knew
now what he had known as soon as he saw Hoke's this morning but had not yet thought into words: why Major de Spain
had not come back, and that after this time he himself, who
had had to see it one time other, would return no more.

                    William Faulkner, "The Bear," 1942 *

THE incursion of the railroad in Sleepy Hollow,
recorded by Hawthorne in 1844, typifies the moment of
discovery. Recall the circumstances. On a fine summer

morning the writer enters the woods and sits down to await "such little events as may happen." Writing in a pleasant if somewhat hackneyed literary idiom, he records sights and sounds. Then, extending his observations to nearby farms and pastures, and to the village in the distance, he sketches an ideal rural scheme. He locates himself at the center of an idyllic domain — a land of order, form, and harmony. Like Virgil's Arcadia or Prospero's "majestic vision" or Jefferson's republic of the middle landscape, this is a self-contained, static world, remote from history, where nature and art are in balance. It is as if the writer had set out to realize in his own person the felicity promised, since Shakespeare's time, by the myth of America as a new beginning.

In its simplest, archetypal form, the myth affirms that Europeans experience a regeneration in the New World. They become new, better, happier men — they are reborn. In most versions the regenerative power is located in the natural terrain: access to undefiled, bountiful, sublime Nature is what accounts for the virtue and special good fortune of Americans. It enables them to design a community in the image of a garden, an ideal fusion of nature with art. The landscape thus becomes the symbolic repository of value of all kinds — economic, political, aesthetic, religious. It has been suggested that the American myth of a new beginning may be a variant of the primal myth described by Joseph Campbell: "a separation from the world, a penetration to some source of power, and a life-enhancing return." Hawthorne's situation in Sleepy Hollow seems to figure a realization of the myth until, suddenly, the harsh whistle of the locomotive fills the air. Then discord replaces harmony and the tranquil mood vanishes. Although he later regains a measure of repose, a sense of loss colors the rest of his notes. His final observation is of some clouds that resemble the "shattered ruins of a dreamer's Utopia." [1]

In spite of the facile resolution and the bland, complaisant tone, there is unmistakable power latent in Hawthorne's casual composition. The sudden appearance of the machine in the garden is an arresting, endlessly evocative image. It causes the instantaneous clash of opposed states of mind: a strong urge to believe in the rural myth along with an awareness of industrialization as counterforce to the myth. Since 1844, this motif has served again and again to order literary experience. It appears everywhere in American writing. In some cases, to be sure, the "little event" is a fictive episode with only vague, incidental symbolic overtones. But in others it is a cardinal metaphor of contradiction, exfoliating, through associated images and ideas, into a design governing the meaning of entire works. In what follows I shall consider three more or less distinct uses, or interpretations, of the Sleepy Hollow motif. For convenience they may be labeled transcendental, tragic, and vernacular. Taken together they exemplify a view of life which dominates much of our literature. It is a complex, distinctively American form of romantic pastoralism.

1

Three months before the episode in Sleepy Hollow, Ralph Waldo Emerson published "The Young American," a version of an address originally delivered in Boston on February 7, 1844. Here Emerson speaks in his public voice as prophet of the American idyll. Combining a vivid, Jeffersonian sense of the land as an economic and political force with a transcendental theory of mind, he expounds what may be called the philosophy of romantic American pastoralism. No major writer has come closer to expressing the popular conception of man's relation to nature in nineteenth-century America.

By the time he composed *Nature* (1836), his first published work of importance, Emerson had adapted the rhetoric of the technological sublime to his own purposes. There, in explaining the use of the outer world, or nature, as commodity, Emerson says that man

> no longer waits for favoring gales, but by means of steam, he realizes the fable of Æolus's bag, and carries the two and thirty winds in the boiler of his boat. To diminish friction, he paves the road with iron bars, and, mounting a coach with a ship-load of men, animals, and merchandise behind him, he darts through the country, from town to town, like an eagle or a swallow through the air. By the aggregate of these aids, how is the face of the world changed, from the era of Noah to that of Napoleon!

In the decade before 1844 Emerson repeatedly draws upon the facts of technological progress to illustrate his exultant sense of human possibilities. For him, new inventions are evidence of man's power to impose his will upon the world. A youthful admirer of Bacon and Franklin, he qualifies but never repudiates the progressive idea of history. He is confident that the advance of science will be followed by a comparable improvement in political morality. A measure of the depth to which machine power penetrates his imagination is the large part it plays in his figurative language. Here, in a later chapter of *Nature*, he is praising intellectual power in general, but, as the submerged metaphor indicates, the specific example he has in mind is technology:

The exercise of the Will, or the lesson of power, is taught in every event. . . . Nature is thoroughly mediate. It is made to serve. . . . It offers all its kingdoms to man as the *raw material* which he may mould into what is useful. Man is never weary of *working it up*. He *forges* the subtle and delicate air into wise and melodious words, and gives them wing as angels of persuasion and command. One after another his victorious thought comes up with and reduces all things, until the world becomes at last only a realized will, — the double of the man. [Italics added.] [2]

It may be surprising, particularly if one thinks of Emerson as a native counterpart of Wordsworth, to find him welcoming the new factory system. One day in 1837, according to his journal, he learns with "sensible relief" that New England is destined to be the "manufacturing country of America." His logic is simple. Now, he explains, better-suited regions will provide food for the nation, and he no longer need feel obliged to sympathize with hard-pressed Yankee farmers. "I am as gay as a canary bird," he says, "with this new knowledge." Piecing together his other comments, we can see that he regards the new technology as an auspicious sign of the times — revolutionary times. It is another outcrop of that international upsurge of energy which had broken through the crust of the old order in the American and French revolutions, and now, half a century later, it is still at work, advancing science and democracy and literature, transforming institutions and, in short, supplanting obsolete forms in every possible sphere of human behavior. A distinct millennial tone may be heard in Emerson's reactions to change in this period. He easily reconciles what often seem in retrospect to have been irreconcilable tendencies, as in this cryptic statement, quoted in its entirety from his journal (1843): "Machinery and Transcendentalism agree well. Stage-Coach and Railroad are bursting the old legislation like green withes." [3]

What perplexes us here is Emerson's ability to join enthusiasm for technological progress with a "romantic" love of nature and contempt for cities. The characteristic starting point of his thought is withdrawal from society in the direction of "nature" — a pastoral impulse. The countryside is the setting of his most intense emotional experiences. "We need nature," he writes, "and cities give the human senses not room enough." It is significant that his literary career begins (in *Nature*) with a pastoral interlude, an ecstatic moment of clarity and joy which comes upon him, unheralded by any special good fortune (like divine and supernatural grace) while crossing a bare common at twilight.* In a sense the rest of Emerson's work is an attempt to construct a coherent philosophy fixed in this moment of integration. Certain that the power to produce such delight does not reside in nature, he tries to determine whether it resides in man or in a harmony of both. Few writers have been more explicit about the supposed metaphysical powers of landscape. In *Nature* Emerson explains that the town common, like the woods, may inspire a return to reason and faith. It is, he says,

---

* The idea of the countryside as the appropriate site of the conversion experience is common to the Christian tradition and the romantic poets. It is the accepted convention in New England Calvinism. Thus Jonathan Edwards describes the sweet sense of God's majesty and grace coming upon him as, in his words, ". . . I walked abroad alone, in a solitary place in my father's pasture, for contemplation." And he describes the state of grace as making the soul "like a field or garden of God, with all manner of pleasant flowers; all pleasant, delightful, and undisturbed; enjoying a sweet calm, and the gently vivifying beams of the sun."

because a more or less untouched landscape provides hints of an "occult relation" with external nature. In such places he feels that he is not alone and unacknowledged. Hence the landscape provides access to first principles. It seems to be a channel of Reason, the higher mode of perception which lends form, value, and meaning to the raw data supplied by the Understanding, the lower mode.[4]

In this reasoning Emerson blends popular American pastoralism with a distinction learned from post-Kantian philosophy. In support of his preference for the "natural" as against the "artificial" (urban) landscape, he invokes the familiar distinction, indirectly derived from Kant, between two faculties of mind: Understanding and Reason. The first is a willed, empirical, practical mode of consciousness (the only reliable instrument of knowledge in the Lockian psychology) which gathers and arranges sense perceptions; the other is the spontaneous, imaginative, mythopoeic, intuitional perception which leaps beyond the evidence of the senses to make analogies and form larger patterns of order. This dualistic theory of mind provides a firm base for romantic pastoralism. Although the prudent, sensible Understanding may be trained in schools and cities, the far-ranging, visionary Reason requires wild or rural scenes for its proper nurture. In such places, says Emerson, reflecting upon his moment of exhilaration on the common, "I find something more dear and connate than in streets or villages." [5]

Nevertheless, and in spite of what he knows about the character of industrial society abroad, Emerson joins in the fervent popular response to the new power. In 1833 he had traveled in England, where he had seen something of the factory system at first hand. For years, moreover, he had been in touch with European literary men; he had

visited Carlyle and Wordsworth, and there can be little
doubt that he appreciated the significance Carlyle attached
to the machine as symbol of an oppressive new order
called industrialism. Yet none of this knowledge dimin-
ishes Emerson's ebullient reaction, in the 1830's and
1840's, to the prospect of steam power, railroads, and
factories in America. He is certain that machinery and
transcendentalism "agree well" — a proposition which he
expounds at length in his 1844 lecture, "The Young
American." [6]

Emerson begins the lecture on a note of patriotic pride.
Taking the rise of a more independent national spirit as
his theme, he announces that America is beginning to
assert herself and that Europe is receding in the native
imagination. At long last a distinctive national culture,
a unique conception of life, is emerging. He attributes
this welcome fact to the combined influence of two forces:
technology and geography — the transportation revolution
and the unspoiled terrain of the new world. Machine
power, which he considers first, is an instrument of
national unity. The new roads, steamboats, and railroads
— especially the railroads — annihilate distance and, "like
enormous shuttles," pattern the various threads of American life into one vast web. As a result, local peculiarities
are overcome, the Union is held staunch, the opening of
the West is accelerated and the influence of Europe weak-
ened. "Railroad iron is a magician's rod," says Emerson,
echoing Samuel Smiles, "in its power to evoke the sleep-
ing energies of land and water." Implicit in this figure
is a popular American attitude toward mechanization.
Like a divining rod, the machine will unearth the hidden
graces of landscape. There are to be no satanic mills in

America, no dark, begrimed cities, nothing like the squalid, inhuman world depicted by Blake, Dickens, and Carlyle.

Certain large features of the society Emerson envisages become apparent when he turns, a moment later, from praise of machines to the denunciation of cities. The new technology is welcome, but cities would come between Americans and the other major source of the rising national spirit, the "bountiful continent," the land itself. He refers to the New World as "our garden," a hemisphere Columbus was encouraged to seek, he says, because the "harmony of nature" required its existence. But what has become of that initial yearning for harmony? As compared with Europe, Emerson is forced to admit, the American scene in 1844 is not pleasing to the eye. It includes few beautiful gardens, either public or private, and the country-side as a whole — land and buildings regarded as one — looks poverty-stricken, plain, and poor. Here Emerson is stating the paradox of American culture in the very terms Robert Beverley had used in 1705. To the Virginian the shabby condition of the landscape had seemed to argue a radical deficiency in the people. He found it difficult to explain why Virginians did not turn their magnificent land into the world's garden. And now Emerson, concerned about the same disturbing fact, contends that the cities are at fault. Cities, he tells the audience of Bostonians, drain the country of the flower of youth, the best part of the population, and leave the countryside (in the absence of a landed aristocracy) to be cultivated by an inferior, irresponsible class. He therefore would arrest the growth of cities, and he urges support of "whatever events," as he puts it, "shall go to disgust men with cities and infuse into them the passion for country life and country pleasures. . . ." One such desirable "event," strangely enough, is the development of machine power. Like Thomas Jefferson, Emerson is confident that under native conditions science and technology can be made to serve a rural ideal.

This confidence rests on what D. H. Lawrence was to call the American spirit of place. Emerson reasons that geography — the incalculable effect of place upon the native consciousness — will make all the difference in the way his countrymen use the new power. He admires machine technology as a superb tool which extends the efficacy of the Understanding. In itself, however, this faculty of practical intelligence is of uncertain value; it is morally neutral, thus potentially dangerous. Men of Understanding are able to manipulate the knowledge won by sense perception, but they are unable to *transcend* it; they perfect the means by which society functions, but they fail to define the ends. A nation dominated by men of this type could become passive in the face of history. As Carlyle had observed, success in dealing with the world as it is may diminish the desire, and the ability, to imagine it as it might be. But Emerson is confident that in Young America mechanical power is to be matched by a new access of vitality to the imaginative, utopian, transcendent, value-creating faculty, Reason. His hope arises from a conviction that men who confront raw nature will ask ultimate questions. He regards the virgin landscape as a source of spiritual therapy, a divine hieroglyph awaiting translation by Americans into aims worthy of their vast new powers. "The land," he explains, "is the appointed remedy for whatever is false and fantastic in our culture. The continent we inhabit is to be physic and food for our mind, as well as our body. The land, with its tranquilizing, sanative influences, is to repair the errors of a scholastic and traditional education, and bring us into just relations with men and things."

Nor is America the only country where the land is to have an increasing influence. Interpreting European politics in the light of native pastoralism, Emerson detects a new ethos, sequel to the capitalist spirit of trade, rising on both sides of the Atlantic. It is making itself felt in various forms of anti-capitalist activity — in the many native experiments in "beneficent socialism," and in the communist movement in France, Germany, and elsewhere. He makes no distinction between mild literary communities like Brook Farm and the gathering forces of violent revolution in Europe. (*The Communist Manifesto* is only four years away.) All of these enterprises, like machinery and transcendentalism, threaten the "old legislation." "These communists," he explains, prefer ". . . the agricultural life as the most favorable condition for human culture. . . ." Initiated by Massachusetts farmers in 1776, the great revolution always had implied renunciation of the city, the traditional home of economic man. In 1844 the revolution is still under way, and what nation is better situated to complete it than Young America? In the end Emerson's rhetoric ascends to Websterian heights:

> Gentlemen, the development of our American internal resources, the extension to the utmost of the commercial system, and the appearance of new moral causes which are to modify the State, are giving an aspect of greatness to the Future, which the imagination fears to open. One thing is plain for all men of common sense and common conscience, that here, here in America, is the home of man. . . . after all the deduction is made for our frivolities and insanities, there still remains an organic sim-

plicity and liberty, which, when it loses its balance, redresses itself presently, which offers opportunity to the human mind not known in any other region.

All in all, there is a striking resemblance between Emerson's Young America and the mythic world inhabited by Jefferson's noble husbandman. (That Jefferson's direct influence upon him was negligible makes the fact even more striking.) Although Emerson attaches more importance to the part played by the new technology, he adopts the Jeffersonian view that it will be redeemed by contact with the virgin land: a "sanative and Americanizing influence, which promises to disclose new virtues for ages to come." The industrial revolution is a railway journey in the direction of nature. "Luckily for us," he says, "now that steam has narrowed the Atlantic to a strait, the nervous, rocky West is intruding a new and continental element into the national mind, and we shall yet have an American genius." The nation is engaged in a pastoral retreat on a monumental scale, and the result is to be a distinctive national culture. Moving west means casting off European attitudes and rigid social forms and urban ways. (The city is an obsolete, quasi-feudal institution.) The machine, appearing at a providential moment, provides access to a "bare common" of continental size, where millions may find a new life. An American genius will arise. He will be saved from satanic mechanism by an influx of grace such as Crèvecœur had described: a total political, economic, and religious conversion. The Young American, like the noble husbandman, will renounce the values of a commercial society. No longer driven by lust for wealth and power, he will adopt material sufficiency as his economic aim. The American faith described by Emerson is a version of pastoral. The renunciation of worldly striving in favor of a simpler, more contemplative life always had been the core of the pastoral ethos; but here, in the New World, the beneficent influence of an unmodified land-scape makes the act credible as never before. "How much better," says Emerson of the movement of men and ma-chines into the West, "when the whole land is a gar-den, and the people have grown up in the bowers of a paradise."

Later, in *The American* (1877), Henry James would lo-cate the essential drama in this native theodicy. At the outset the hero, still a young man, describes his renuncia-tion of a successful business career. He had come to New York on a "big transaction," where he stood to get his own back ("a remarkably sweet thing, a matter of half a mil-lion") from a party who had wronged him. But then a curious thing had happened. He had fallen asleep while riding in a hack and, as he explains,

> ". . . I woke up suddenly, from a sleep or from a kind of reverie, with the most extraordinary change of heart — a moral disgust for the whole proposition. It came upon me like *that!*" — and he snapped his fingers — "as abruptly as an old wound that begins to ache. I could n't tell the meaning of it; I only realised I had turned against my-self worse than against the man I wanted to smash. The idea of *not* coming by that half-million in that particular way, of letting it utterly slide and scuttle and never hear-ing of it again, became the one thing to save my life from a sudden danger. And all this took place quite in-dependently of my will . . ."

His companion is incredulous. Had he simply allowed the other man to collar the half-million? James's hero does not

know. When the hack reached its destination, he had realized that he could not move. " 'What I wanted to get out
of,' " he says, " 'was Wall Street.' "

> "I told the man to drive to the Brooklyn ferry and cross
> over. When we were over I told him to drive me out into
> the country. As I had told him originally to drive for dear
> life down town, I suppose he thought I had lost my wits
> on the way. Perhaps I had, but in that case my sacrifice
> of them has become, in another way, my biggest stroke
> of business."

It is at this point that James explains the American
hero's allegorical name: " 'I spent the morning,' " he
says, " 'looking at the first green leaves on Long Island.
. . . As for the damned old money, I've enough, already,
not to miss it. . . . I seemed to feel a new man under my
old skin; at all events I longed for a new world.' " The
germ of irony, later to develop into the "international
theme," is that Christopher Newman's "new world" is
Europe.[7]

But to return to Emerson. Even in moments of extravagant hope, he did not imagine that the conversion
could be accomplished without effort. Besides exhortations
to the public like "The Young American," he issued calls
to scholars, artists, and writers, urging them to lead the
way. Their duty, as he saw it, was to exemplify the subordination of power to purpose. Almost a century before
Hart Crane's pronouncements on the subject — "unless
poetry can absorb the machine," said Crane, "i.e., *ac-
climatize* it as naturally and casually as trees, cattle, gal-
leons, castles . . . [it] has failed of its full contemporary
function" — Emerson recognized the analogy between the
political and aesthetic problems of the age. In his essays
on "Art" (1841) and "The Poet" (1844) he urged Ameri-

can intellectuals to conquer the new territory being opened up by industrialization. The two essays provide an aesthetic corollary to the program of "The Young American." [8]

Readers of poetry, says Emerson, "see the factory-village and the railway, and fancy that the poetry of the landscape is broken up by these," but they are mistaken. They assume that the new technology is anti-poetic, because, for one thing, "these works of art are not yet consecrated in their reading." But there is nothing inherently ugly about factories and railroads; what is ugly is the dislocation and detachment from "the Whole" which they represent when seen only from the limited perspective of the Understanding. The literary problem, accordingly, is society's problem in small. To dispel the ugliness which surrounds the new technology, whether in a poem or a landscape, we must assign it to its proper place in the human scale. Artists have a special responsibility to incorporate into their work such "new and necessary facts" as the shop, mill, and railroad. For there will always be new things under the sun, and it requires genius to reveal their beauty and value. Nothing is anti-poetic; thought "makes everything fit for use." The poet, "who re-attaches things to nature and the Whole, — re-attaching even artificial things and violations of nature, to nature, by a deeper insight, — disposes very easily of the most disagreeable facts." Blessed with the high perception, he sees the factory town and railway "fall within the great Order not less than the beehive or the spider's geometrical web. Nature adopts them very fast into her vital circles, and the gliding train of cars she loves like her own."

Emerson's attitude follows from the assumptions of philosophic idealism. What we do with instruments of power depends, finally, upon the way we perceive them. His position is like the one Shakespeare had assigned to Polixenes in the pastoral episode of *The Winter's Tale.* The conflict between art and nature must be reconcilable because:

> . . . Nature is made better by no mean
> But Nature makes that mean; so, over that art
> Which you say adds to Nature, is an art
> That Nature makes.

If technology is the creation of man, who is a product of nature, then how can the machine in the landscape be thought to represent an unresolvable conflict? Men of genius, who perceive relations hidden from other men, disclose the underlying unity of experience and so help to direct the course of events. A great poet not only asserts but exemplifies the possibility of harmony. When he assimilates new and seemingly artificial facts into the texture of a poem, he provides an example for all men. What he achieves in art they can achieve in life.

2

Soon after Emerson had set forth his program for Young Americans, his young disciple Henry Thoreau put it to a test. He began his stay at Walden Pond in the spring of 1845, and the book he eventually wrote about it (*Walden* was not published until 1854) may be read as the report of an experiment in transcendental pastoralism. The organizing design is like that of many American fables: *Walden* begins with the hero's withdrawal from society in the direction of nature. The main portion of the book is given over to a yearlong trial of Emerson's prescription for achieving a new life. When Thoreau tells of his return to Concord, in the end, he seems to have satisfied himself about the efficacy of this method of redemption. It may be difficult to say exactly what is being claimed, but the triumphant tone of the concluding chapters leaves little doubt that he is announcing positive results. His most telling piece of evidence is *Walden* — the book itself. Recognizing the clarity, coherence, and power of the writing, we can only conclude — or so transcendental doctrine would have it — that the experiment has been a success. The vision of unity that had made the aesthetic order of *Walden* possible had in turn been made possible by the retreat to the pond. The pastoral impulse somehow had provided access to the order latent in the cosmos.

But the meaning of *Walden* is more complicated than this affirmation. Because Thoreau takes seriously what Emerson calls the "method of nature" — more seriously than the master himself — the book has a strong contrapuntal theme. Assuming that natural facts properly perceived and accurately transcribed must yield the truth, Thoreau adopts the tone of a hard-headed empiricist. At the outset he makes it clear that he will tell exactly what happened. He claims to have a craving for reality (be it life or death), and he would have us believe him capable of reporting the negative evidence. Again and again he allows the facts to play against his desire, so that his prose at its best acquires a distinctively firm, cross-grained texture. Though the dominant tone is affirmative, the undertone is skeptical, and it qualifies the import of episode after episode. For this reason *Walden* belongs among the first in a long series of American books which, taken together, have had the effect of circumscribing the pastoral hope, much as Virgil circumscribes it in his eclogues. In form and feeling, indeed, Thoreau's book has much in common with the classic Virgilian mode.

Although the evidence is abundant, it is easy to miss the conventional aspect of *Walden*. In the second chapter Thoreau describes the site as an ideal pasture, a real place which he transforms into an unbounded, timeless landscape of the mind. And he identifies himself with Damodara (Krishna) in his rôle as shepherd, and with a shepherd in a Jacobean song:

> There was a shepherd that did live,
>   And held his thoughts as high
> As were the mounts whereon his flocks
>   Did hourly feed him by.

Nevertheless, the serious affinity between *Walden* and the convention is disguised by certain peculiarities of American pastoralism, the most obvious being the literalness with which Thoreau approaches the ideal of the simple life. For centuries writers working in the mode had been playing with the theme, suggesting that men might enrich their contemplative experience by simplifying their housekeeping. (The shepherd's ability to reduce his material needs to a minimum had been one of his endearing traits.) Yet it generally had been assumed that the simple life was a poetic theme, not to be confused with the way poets did in fact live. In the main, writers who took the felicity of shepherds in green pastures as their subject had been careful to situate themselves near wealth and power. The effect of the American environment, however, was to break down common-sense distinctions between art and life. No one understood this more clearly than Henry

Thoreau; skilled in the national art of disguising art, in *Walden* he succeeds in obscuring the traditional, literary character of the pastoral withdrawal. Instead of writing about it — or *merely* writing about it — he tries it. By telling his tale in the first person, he endows the mode with a credibility it had seldom, if ever, possessed. Because the "I" who addresses us in *Walden* is describing the way he had lived, taking pains to supply plenty of hard facts ("Yes, I did eat $8.74, all told. . . ."), we scarcely notice that all the while he had been playing the shepherd's venerable rôle. He refuses to say whether the book is an explicit guide for living or an exercise in imaginative perception. We are invited to take it as either or both. Convinced that effective symbols can be derived only from natural facts, Thoreau had moved to the pond so that he might make a symbol of his life. If we miss the affinity with the Virgilian mode, then, it is partly because we are dealing with a distinctively American version of romantic pastoral.[9]

No feature of *Walden* makes this truth more apparent than its topography. The seemingly realistic setting may not be a land of fantasy like Arcadia, yet neither is it Massachusetts. On inspection it proves to be another embodiment of the American moral geography — a native blend of myth and reality. The hut beside the pond stands at the center of a symbolic landscape in which the village of Concord appears on one side and a vast reach of unmodified nature on the other. As if no organized society existed to the west, the mysterious, untrammeled, primal world seems to begin at the village limits. As in most American fables, the wilderness is an indispensable feature of this terrain, and the hero's initial recoil from everyday life carries him to the verge of anarchic primi-

tivism.* "We need the tonic of wildness," Thoreau explains, using the word "pasture" to encompass wild nature: "We need to witness our own limits transgressed,
and some life pasturing freely where we never wander."
(The combined influence of geography and the romantic
idea of nature — sublime Nature — gives rise to attitudes
held by a long line of American literary heroes from
Natty Bumppo to Ike McCaslin.) But Thoreau is not a
primitivist. True, he implies that he would have no difficulty choosing between Concord and the wilderness.
What really engages him, however, is the possibility of
avoiding that choice. (Jefferson had taken the same position.) In *Walden*, accordingly, he keeps our attention
focused upon the middle ground where he builds a house,
raises beans, reads the *Iliad*, and searches the depths of
the pond. Like the "navel of the earth" in the archaic
myths studied by Mircea Eliade, the pond is the absolute
center — the *axis mundi* — of Thoreau's cosmos. If an
alternative to the ways of Concord is to be found anywhere, it will be found on the shore of Walden Pond —
near the mystic center.[10]

And it had best be found quickly. The drama of *Walden*
is intensified by Thoreau's acute sense of having been
born in the nick of time. Though the book resembles the
classic pastoral in form and feeling, its facts and images

* The difference between the typical American hero and the shepherd
in traditional versions of pastoral is suggested by Renato Poggioli's account of that archetypal figure as one who "lives a sedentary life even in
the open, since he prefers to linger in a grove's shade rather than to
wander in the woods. He never confronts the true wild, and this is why
he never becomes even a part-time hunter." Given the circumstances of
American life, our heroes do confront the true wild, and they often become hunters. But it is striking to notice how often they are impelled
to restrict or even renounce their hunting. I am thinking of Natty
Bumppo, Melville's Ishmael, Faulkner's Ike McCaslin, and Thoreau himself.

are drawn from the circumstances of life in nineteenth-century America. By 1845, according to Thoreau, a depressing state of mind — he calls it "quiet desperation" — has seized the people of Concord. The opening chapter, "Economy," is a diagnosis of this cultural malady. Resigned to a pointless, dull, routinized existence, Thoreau's fellow-townsmen perform the daily round without joy or anger or genuine exercise of will. As if their minds were mirrors, able only to reflect the external world, they are satisfied to cope with things as they are. In Emerson's language, they live wholly on the plane of the Understanding. Rather than design houses to fulfill the purpose of their lives, they accommodate their lives to the standard design of houses. Thoreau discovers the same pattern of acquiescence, a dehumanizing reversal of ends and means, in all of their behavior. He finds it in their pretentious furnishings, their uncomfortable clothing, their grim factories, the dispirited way they eat and farm the land and work from dawn to dusk. He locates it, above all, in their economy — a system within which they work endlessly, not to reach a goal of their own choosing but to satisfy the demands of the market mechanism. The moral, in short, is that here "men have become the tools of their tools."

The omnipresence of tools, gadgets, instruments is symptomatic of the Concord way. Like Carlyle, Thoreau uses technological imagery to represent more than industrialization in the narrow, economic sense. It accompanies a mode of perception, an emergent system of meaning and value — a culture. In fact his overdrawn indictment of the Concord "economy" might have been written to document Carlyle's dark view of industrialism. Thoreau feels no simple-minded Luddite hostility toward the new inventions; they are, he says, "but improved means to an unimproved end. . . ." What he is attacking is the popular illusion that improving the means is enough, that if the machinery of society is put in good order (as Carlyle had said) "all were well with us; the rest would care for itself!" He is contending against a culture pervaded by this mechanistic outlook. It may well be conducive to material progress, but it also engenders deadly fatalism and despair. At the outset, then, Thoreau invokes the image of the machine to represent the whole tone and quality of Concord life or, to be more precise, anti-life:

> Actually, the laboring man has not leisure for a true integrity day by day; he cannot afford to sustain the manliest relations to men; his labor would be depreciated in the market. He has no time to be anything but a machine.

The clock, favorite "machine" of the Enlightenment, is a master machine in Thoreau's model of the capitalist economy. Its function is decisive because it links the industrial apparatus with consciousness. The laboring man becomes a machine in the sense that his life becomes more closely geared to an impersonal and seemingly autonomous system.* If the advent of power technology is alarming, it is because it occurs within this cultural context. When Thoreau depicts the machine as it functions within the

---

* Thoreau's response to the mechanization of time reflects the heightened significance of the clock in the period of the "take-off" into full-scale industrialism. With the building of factories and railroads it became necessary, as never before, to provide the population with access to the exact time. This was made possible, in New England, by the transformation of the clockmaking industry. Before 1800 clocks had been relatively expensive luxury items made only by master craftsmen. Significantly enough, the industry was among the first to use machines and the principle of interchangeable part manufacture. By 1807, in Connecticut, Eli Terry had begun to produce wooden clocks in large numbers, and before he died in 1852 he was making between 10,000 and 12,000 clocks a year sold at $5.00 each.

Concord environment, accordingly, it is an instrument of oppression: "We do not ride upon the railroad; it rides upon us." But later, when seen from the Walden perspective, the railroad's significance becomes quite different.[11]

Thoreau's denunciation of the Concord "economy" prefigures the complex version of the Sleepy Hollow episode in the fourth chapter, "Sounds." The previous chapter is about "Reading," or what he calls the language of metaphor. Now he shifts to sounds, "the language which all things and events speak without metaphor, which alone is copious and standard." The implication is that he is turning from the conventional language of art to the spontaneous language of nature. What concerns him is the hope of making the word one with the thing, the notion that the naked fact of sensation, if described with sufficient precision, can be made to yield its secret — its absolute meaning. This is another way of talking about the capacity of nature to "produce delight" — to supply value and meaning. It is the crux of transcendental pastoralism. Hence Thoreau begins with an account of magnificent summer days when, like Hawthorne at the Hollow, he does nothing but sit "rapt in a revery, amidst the pines and . . . sumachs, in undisturbed solitude and stillness." These days, unlike days in Concord, are not "minced into hours and fretted by the ticking of a clock." Here is another pastoral interlude, a celebration of idleness and that sense of relaxed solidarity with the universe that presumably comes with close attention to the language of nature. For a moment Thoreau allows us to imagine that he has escaped the clock, the Concord definition of time and, indeed, the dominion of the machine. But then, without raising his voice, he reports the "rattle of railroad cars" in the woods.

At first the sound is scarcely audible. Thoreau casually

mentions it at the end of a long sentence in which he describes a series of sights and sounds: hawks circling the clearing, a tantivy of wild pigeons, a mink stealing out of the marsh, the sedge bending under the weight of reed-birds, and then, as if belonging to the very tissue of nature: "and for the last half-hour I have heard the rattle of railroad cars, now dying away and then reviving like the beat of a partridge, conveying travellers from Boston to the country." It would have been difficult to contrive a quieter entrance, which may seem curious in view of the fact that Thoreau then devotes nine long paragraphs to the subject. Besides, he insists upon the importance of the Fitchburg Railroad in the Walden scene; it "touches the pond" near his house, and since he usually goes to the village along its causeway, he says, "I . . . am, as it were, related to society by this link." * And then, what may at first seem even more curious, he introduces the auditory image of the train a second time, and with a markedly different emphasis:

> The whistle of the locomotive penetrates my woods summer and winter, sounding like the scream of a hawk sailing over some farmer's yard, informing me that many restless city merchants are arriving within the circle of the town. . . .

Now the sound is more like a hawk than a partridge, and Thoreau playfully associates the hawk's rapacity with the train's distinctive mechanical cadence:

> All the Indian huckleberry hills are stripped, all the cranberry meadows are raked into the city. Up comes

* It is significant that Thoreau added this statement, with its obvious claim for the symbolic significance of the railroad, to the version of the episode he had published earlier in *Sartain's Union Magazine*, XI (1852), 66–8.

the cotton, down goes the woven cloth; up comes the silk, down goes the woollen; up come the books, but down goes the wit that writes them.

What are we to make of this double image of the railroad? First it is like a partridge, then a hawk; first it blends into the landscape like the industrial images in the Inness painting, but then, a moment later, it becomes the discordant machine of the Sleepy Hollow notes. What does the railroad signify here? On inspection the passage proves to be a sustained evocation of the ambiguous meaning of the machine and its relation to nature. Every significant image is yoked to an alternate:

> When I meet the engine with its train of cars moving off with planetary motion,— or, rather, like a comet . . .

Or the cloud of smoke

> . . . rising higher and higher, going to heaven while the cars are going to Boston, conceals the sun for a minute and casts my distant field into the shade. . . .

The point becomes explicit in a thought that Thoreau repeats like a refrain: "If all were as it seems, and men made the elements their servants for noble ends!"

The image of the railroad on the shore of the pond figures an ambiguity at the heart of *Walden*. Man-made power, the machine with its fire, smoke, and thunder, is juxtaposed to the waters of Walden, remarkable for their depth and purity and a matchless, indescribable color — now light blue, now green, almost always pellucid. The iron horse moves across the surface of the earth; the pond invites the eye below the surface. The contrast embodies both the hope and the fear aroused by the impending climax of America's encounter with wild nature. As

Thoreau describes the event, both responses are plausible, and there is no way of knowing which of them history is more likely to confirm. Earlier he had made plain the danger of technological progress, and here at the pond it again distracts his attention from other, presumably more important, concerns. Yet he is elated by the presence of this wonderful invention. In Concord, within the dominion of the mechanistic philosophy, the machine rode upon men, but when seen undistorted from Walden, the promise of the new power seems to offset the danger. Thoreau is delighted by the electric atmosphere of the depot and the cheerful valor of the snow-plow crews. He admires the punctuality, the urge toward precision and order, the confidence, serenity, and adventurousness of the men who operate this commercial enterprise:

> . . . when I hear the iron horse make the hills echo with his snort like thunder, shaking the earth with his feet, and breathing fire and smoke from his nostrils (what kind of winged horse or fiery dragon they will put into the new Mythology I don't know), it seems as if the earth had got a race now worthy to inhabit it. If all were as it seems, and men made the elements their servants for noble ends!

If the interrupted idyll represents a crucial ambiguity, it also represents at least one certainty. The certainty is change itself — the kind of accelerating change, or "progress," that Americans identify with their new inventions, especially the railroad. For Thoreau, like Melville's Ahab, this machine is the type and agent of an irreversible process: not mere scientific or technological development in the narrow sense, but the implacable advance of history. "We have constructed a fate," he writes, "an *Atropos*, that never turns aside. (Let that be the name of your engine.)" The episode demonstrates that the Walden site cannot provide a refuge, in any literal sense, from the forces of change. Indeed, the presence of the machine in the woods casts a shadow of doubt (the smoke of the locomotive puts Thoreau's field in the shade) upon the Emersonian hope of extracting an answer from nature. The doubt is implicit in the elaborately contrived language used to compose this little event. Recall that Thoreau had introduced the chapter on "Sounds" as an effort to wrest an extra-literary meaning from natural facts; his alleged aim had been to render sense perceptions with perfect precision in "the language which all things and events speak without metaphor." What he actually had done, however, was quite the reverse. To convey his response to the sound of the railroad he had resorted to an unmistakably figurative, literary language. Few passages in *Walden* are more transparently contrived or artful; it is as if the subject had compelled Thoreau to admit a debt to Art as great, if not greater, than his debt to Nature.

The most telling qualification of Emersonian optimism, however, comes in the deceptively plain-spoken conclusion to the episode. Emerson had affirmed the political as well as the religious value of the pastoral impulse. When he spoke in his public voice (as in "The Young American") he interpreted the nation's movement toward "nature" (signifying both a natural and a spiritual fact — both land and landscape) as motion toward a new kind of technically advanced yet rural society. In effect he was reaffirming the Jeffersonian hope of embodying the pastoral dream in social institutions. But Thoreau, abiding by his commitment to stand "right fronting and face to face to a fact," takes another hard look at the sight of the machine in the American landscape:

And hark! here comes the cattle-train bearing the cattle of a thousand hills, sheepcots, stables, and cow-yards in the air, drovers with their sticks, and shepherd boys in the midst of their flocks, all but the mountain pastures, whirled along like leaves blown from the mountains by the September gales. The air is filled with the bleating of calves and sheep, and the hustling of oxen, as if a pastoral valley were going by. . . . A carload of drovers, too, in the midst, on a level with their droves now, their vocation gone, but still clinging to their useless sticks as their badge of office. . . . So is your pastoral life whirled past and away. But the bell rings, and I must get off the track and let the cars go by; —

> What's the railroad to me?
> I never go to see
> Where it ends.
> It fills a few hollows,
> And makes banks for the swallows,
> It sets the sand a-blowing,
> And the blackberries a-growing.

but I cross it like a cart-path in the woods. I will not have my eyes put out and my ears spoiled by its smoke and steam and hissing.

Compared to popular, sentimental pastoralism, or to Emerson's well-turned evasions, there is a pleasing freshness about Thoreau's cool clarity. He says that the pastoral way of life — pastoralism in the literal, agrarian sense — is being whirled past and away. It is doomed. And he has no use for the illusion that the *Atropos* can be stopped. The first thing to do, then, the only sensible thing to do, is get off the track. Not that one need resign oneself, like the men of Concord, to the dominion of the mechanical philosophy. But how is the alternative to be

defined? To answer the question had been the initial purpose of the Walden experiment; now its urgency is heightened by the incursion of history. If he is to find an answer,
the writer's first duty is to protect his powers of perception.
At this point Thoreau adopts a testy, tight-lipped, un-
compromising tone: "I will not have my eyes put out and
my ears spoiled by its smoke and steam and hissing."

The need for defense against the forces of history does
not tempt Thoreau to a nostalgic embrace of the "pastoral
life" that is being whirled away. Quite the contrary. In
"The Bean-Field" he turns his wit against the popular
American version of pastoral. The Walden experiment,
as described in "Economy," had included a venture in
commercial farming. In order to earn ten or twelve dollars by an "honest and agreeable method," he had planted
two acres and a half, chiefly with beans. That is a lot of
beans (he figures that the length of the rows, added together, was seven miles), and it meant a lot of work. Here,
then, in "The Bean-Field" he turns to the "meaning of
this so steady and self-respecting, this small Herculean labor. . . ." The chapter is a seriocomic effort to get at
the lesson of agricultural experience. At the outset, recalling his first visit to the pond as a child, Thoreau invests the scene of his arduous labor with an appropriate
bucolic ambience:

> And now to-night my flute has waked the echoes over
> that very water. . . . Almost the same johnswort springs
> from the same perennial root in this pasture, and even
> I have at length helped to clothe that fabulous land-
> scape of my infant dreams, and one of the results of my
> presence and influence is seen in these bean leaves. . . .

As he describes himself at work among his beans,
Thoreau is the American husbandman. Like the central

figure of the Jeffersonian idyll, his vocation has a moral and spiritual as well as economic significance. And his field bears a special relation to American circumstances. It produces beans which resemble neither English hay, with its synthetic quality (a result of precise, calculating, scientific methods), nor the rich and various crop produced spontaneously in the surrounding woods, pastures, and swamps. "Mine was, as it were, the connecting link between wild and cultivated fields; as some states are civilized, and others half-civilized, and others savage or barbarous, so my field was, though not in a bad sense, a half-cultivated field." But Thoreau's husbandman cannot be characterized simply by his location in the middle landscape. Like Emerson's Young American, he blends Jeffersonian and romantic attitudes toward nature. When Thoreau describes the purpose of his bean-raising activity, accordingly, he falls into a comic idiom — a strange compound of practical, Yankee vernacular and transcendental philosophizing:

> It was a singular experience that long acquaintance which I cultivated with beans, what with planting, and hoeing, and harvesting, and threshing, and picking over and selling them, — the last was the hardest of all, — I might add eating, for I did taste. I was determined to know beans.

The better he had come to "know beans," the less seriously he had been able to take the rôle of noble husbandman. As the writer's account of that "singular experience" develops, he moves further and further from the reverential, solemn tone of popular pastoralism, until he finally adopts a mock-heroic attitude:

> I was determined to know beans. When they were growing, I used to hoe from five o'clock in the morning till

noon, and commonly spent the rest of the day about other affairs. Consider the intimate and curious acquaintance one makes with various kinds of weeds,—it will bear some iteration in the account, for there was no little iteration in the labor,—disturbing their delicate organizations so ruthlessly, and making such invidious distinctions with his hoe, levelling whole ranks of one species, and sedulously cultivating another. That's Roman wormwood,—that's pigweed,—that's sorrel,—that's piper-grass, —have at him, chop him up, turn his roots upward to the sun, don't let him have a fibre in the shade, if you do he'll turn himself t'other side up and be as green as a leek in two days. A long war, not with cranes, but with weeds, those Trojans who had sun and rain and dews on their side. Daily the beans saw me come to their rescue armed with a hoe, and thin the ranks of their enemies, filling up the trenches with weedy dead. Many a lusty crest-waving Hector, that towered a whole foot above his crowding comrades, fell before my weapon and rolled in the dust.

In part Thoreau's irony can be attributed to the outcome of his bean venture. Although he does not call it a failure, the fact is clear enough. The cost of the operation was $14.72½ (he had hired a man to help with the plowing), the gross income $23.44, and the net profit $8.71½. This sum barely paid for the rest of his food, and to make ends meet he had hired himself out as a day laborer. In other words, the bean crop did not provide an adequate economic base for the life of an independent husbandman. His own experience comports with what he observes of American farmers throughout the book. So far from representing a "pastoral life," a desirable alternative to the ways of Concord and the market economy, the typical farmer in *Walden* is narrow-minded and greedy. The description of the ice crews stripping the pond in winter, as if they were hooking up "the virgin mould itself," is a bitter comment on the methods of capitalist "husband-men." Thoreau has no use for the cant about the nobility of the farmer. ("I respect not his labors, his farm where everything has its price, who would carry the landscape, who would carry his God, to market, if he could get anything for him; who goes to market *for* his god as it is. . . .") In concluding "The Bean-Field" he strips native agriculture of its meretricious rhetorical disguise:

> Ancient poetry and mythology suggest, at least, that husbandry was once a sacred art; but it is pursued with irreverent haste and heedlessness by us, our object being to have large farms and large crops merely. We have no festival, nor procession, nor ceremony, not excepting our cattle-shows and so-called Thanksgivings, by which the farmer expresses a sense of the sacredness of his calling, or is reminded of its sacred origin. It is the premium and the feast which tempt him. He sacrifices not to Ceres and the Terrestrial Jove, but to the infernal Plutus rather. By avarice and selfishness, and a grovelling habit, from which none of us is free, of regarding the soil as property, or the means of acquiring property chiefly, the landscape is deformed, husbandry is degraded with us, and the farmer leads the meanest of lives. He knows Nature but as a robber.[12]

The result of the venture in husbandry prefigures the result of the Walden experiment as a whole. Judged by a conventional (economic) standard, it is true, the enterprise had been a failure. But that judgment is irrelevant to Thoreau's purpose, as his dominant tone, the tone of success, plainly indicates. It is irrelevant because his aim had been to *know* beans: to get at the essential *meaning* of labor in the bean-field. And "meaning," as he conceives it, has nothing to do with the alleged virtue of the American husbandman or the merits of any institution or "way of life"; nor can it be located in the material or economic facts, where Concord, operating on the plane of the Understanding, locates meaning and value. Thoreau has quite another sort of meaning in view, as he admits when he says that he raised beans, not because he wanted beans to eat, "but, perchance, as some must work in fields if only for the sake of tropes and expression, to serve a parable-maker one day."

This idea, which contains the gist of Thoreau's ultimate argument, also is implicit in the outcome of other episodes. It is implied by his account of fishing at night — a tantalizing effort to get at the "dull uncertain blundering purpose" he detects at the end of his line beneath the pond's opaque surface; and by that incomparable satire on the transcendental quest, the chase of the loon who "laughed in derision" at his efforts; and by his painstaking investigation of the pond's supposed "bottomlessness": ". . . I can assure my readers that Walden has a reasonably tight bottom at a not unreasonable . . . depth." In each case, as in "The Bean-Field," the bare, empirical evidence proves inadequate to his purpose. Of themselves the facts do not, cannot, flower into truth; they do not show forth a meaning, which is to say, the kind of meaning the experiment had been designed to establish. If the promise of romantic pastoralism is to be fulfilled, nothing less than an alternative to the Concord way will suffice. Although his tone generally is confident, Thoreau cunningly keeps the issue in doubt until the end. By cheerfully, enigmatically reiterating his failure to extract an "answer" — a coherent world-view — from the facts, he moves the drama toward a climax. Not until the penultimate chapter, "Spring," does he disclose a way of coping with the forces represented by the encroaching machine power.[13]

At the same time, however, he carefully nurtures an awareness of the railroad's presence in the Concord woods. (The account of the interrupted idyll in "Sounds" is only the most dramatic of its many appearances.) There is scarcely a chapter in which he does not mention seeing or hearing the engine, or walking "over the long causeway made for the railroad through the meadows. . . ." When the crew arrives to strip the ice from the pond, it is "with a peculiar shriek from the locomotive." And Thoreau takes special pains to impress us with the "cut" in the landscape made by the embankment. He introduces the motif in the first chapter, after describing his initial visit to the Walden site:

> . . . I came out on to the railroad, on my way home, its yellow sand-heap stretched away gleaming in the hazy atmosphere, and the rails shone in the spring sun . . .

And he returns to it in "The Ponds":

> That devilish Iron Horse, whose ear-rending neigh is heard throughout the town, has muddied the Boiling Spring with his foot, and he it is that has browsed off all the woods on Walden shore, that Trojan horse, with a thousand men in his belly, introduced by mercenary Greeks! Where is the country's champion, the Moore of Moore Hall, to meet him at the Deep Cut and thrust an avenging lance between the ribs of the bloated pest?

The Deep Cut is a wound inflicted upon the land by man's meddling, aggressive, rational intellect, and it is not healed until the book's climax, the resurgence of life in "Spring." By that point the organizing design of

Walden has been made to conform to the design of nature itself; like Spenser's arrangement of his eclogues in *The Shepheards Calendar,* the sequence of Thoreau's final chapters follows the sequence of months and seasons. This device affirms the possibility of redemption from time, the movement away from Concord time, defined by the clock, toward nature's time, the daily and seasonal life cycle. It is also the movement that redeems machine power. In the spring the ice, sand, and clay of the railroad causeway thaws. The wet stuff flows down the banks, assumes myriad forms, and arouses in Thoreau a delight approaching religious ecstasy. The event provides this parable-maker with his climactic trope: a visual image that figures the realization of the pastoral ideal in the age of machines.

The description of the melting railroad bank is an intricately orchestrated paean to the power of the imagination. Although the sand remains mere sand, the warming influence of the sun causes it to assume forms like lava, leaves, vines, coral, leopards' paws, birds' feet, stalactites, blood vessels, brains, bowels, and excrement. It is a pageant evoking the birth of life out of inorganic matter. Watching the sandy rupture exhilarates Thoreau, affecting him as if, he says, "I stood in the laboratory of the Artist who made the world and me, — had come to where he was still at work, sporting on this bank, and with excess of energy strewing his fresh designs about." The scene illustrates the principle in all the operations of Nature: an urge toward organization, form, design. Every detail confirms the endless creation of new forms. Out of winter's frost comes spring; out of an excrementous flow, newborn creatures; out of the landscape eroded by men and machines, these forms of the molten earth. "There is nothing inorganic." The sight inspires Thoreau with a sense of infinite possibility. "The very globe continually transcends and translates itself. . . ." And not only the earth, he says, "but the institutions upon it are plastic like clay in the hands of the potter."

Thoreau's study of the melting bank is a figurative restoration of the form and unity severed by the mechanized forces of history. Out of the ugly "cut" in the landscape he fashions an image of a new beginning. Order, form, and meaning are restored, but it is a blatantly, unequivocally figurative restoration. The whole force of the passage arises from its extravagantly metaphoric, poetic, literary character. At no point does Thoreau impute material reality to the notion of sand being transformed into, say, leopards' paws. It assumes a form that looks like leopards' paws, but the form exists only so far as it is perceived. The same may be said of his alternative to the Concord way. Shortly after the episode of the thawing sand, the account of the coming of spring reaches a moment of "seemingly instantaneous" change. A sudden influx of light fills his house; he looks out of the window, and where the day before there had been cold gray ice there lies the calm transparent pond; he hears a robin singing in the distance and honking geese flying low over the woods. It is spring. Its coming, says Thoreau, is "like the creation of Cosmos out of Chaos and the realization of the Golden Age."

This reaffirmation of the pastoral ideal is not at all like Emerson's prophecy, in "The Young American," of a time "when the whole land is a garden, and the people have grown up in the bowers of a paradise." By comparison, the findings of the Walden experiment seem the work of a tough, unillusioned empiricist.* They are consistent with Thoreau's unsparing analysis of the Concord "economy" and with the knowledge that industrial prog-

---

* In fairness to Emerson it should be said that by the late 1840's he, too, had become more skeptical about the compatibility of the pastoral ideal and industrial progress. His second visit to England in 1847 was in many ways a turning point in his intellectual development, and *English Traits* (1856) is one of our first and most penetrating studies of the new culture of industrialism. Ostensibly about England, the book manifestly was written with America's future economic development in view. Both the structure and the content of the book are governed by the machine-in-the-garden figure. "England is a garden," says Emerson in "Land" (Ch. 3), and in the first half of the book the rising power of Britain is treated affirmatively. But the argument pivots on "Wealth," the tenth of nineteen chapters, and "Wealth" itself is divided in two. It begins with a tribute to the industrial revolution, to that "marvellous machinery which differences this age from any other age." Both the chapter and the book "break," as it were, on the theme "the machine unmans the user." Machinery, Emerson argues, has proved unmanageable. "Steam from the first hissed and screamed to warn him; it was dreadful with its explosion, and crushed the engineer." He therefore turns, in the second half, to an account of the "civility of trifles, of money and expense, an erudition of sensation [that] takes place, and the putting of as many impediments as . . . [possible] between the man and his objects." In the second half of *English Traits* technological development is a cause of alienation. In the chapter "Literature" (Ch. 14), Emerson finally attempts to locate the source of England's failure to realize its promise. He identifies the "influx of decomposition" with John Locke. Bacon, who "held of the analogists," he says, had been "capable of ideas, yet devoted to ends." But Locke was the exemplar of a "so-called" scientific tendency, a negative and poisonous mode of thought. Like Carlyle, Emerson sees the empirical philosophy as the root of modern powerlessness. The book ends with a speech in which Emerson announces his confidence in Britain's capacity to regain control of her economic affairs. But the final sentences strike this characteristically American chord: "If it be not so, if the courage of England goes with the chances of a commercial crisis, I will go back to the capes of Massachusetts and my own Indian stream, and say to my countrymen, the old race are all gone, and the elasticity and hope of mankind must henceforth remain on the Alleghany ranges, or nowhere."

ress is making nonsense of the popular notion of a "pastoral life." The melting of the bank and the coming of spring is only "like" a realization of the golden age. It is a poetic figure. In *Walden* Thoreau is clear, as Emerson seldom was, about the location of meaning and value. He is saying that it does not reside in the natural facts or in social institutions or in anything "out there," but in consciousness. It is a product of imaginative perception, of the analogy-perceiving, metaphor-making, mythopoeic power of the human mind. For Thoreau the realization of the golden age is, finally, a matter of private and, in fact, literary experience. Since it has nothing to do with the environment, with social institutions or material reality (any facts will melt if the heat of imaginative passion is sufficient), then the writer's physical location is of no great moment. At the end of the chapter on "Spring," accordingly, Thoreau suddenly drops the language of metaphor and reverts to a direct, matter-of-fact, referential idiom: "Thus was my first year's life in the woods completed; and the second year was similar to it. I finally left Walden September 6th, 1847."

There is a world of meaning in the casual tone. If the book ended here, indeed, one might conclude that Thoreau, like Prospero at the end of *The Tempest*, was absolutely confident about his impending return to society. (Concord is the Milan of *Walden*.) But the book does not end with "Spring." Thoreau finds it necessary to add a didactic conclusion, as if he did not fully trust the power of metaphor after all. And he betrays his uneasiness, finally, in the arrogance with which he announces his disdain for the common life:

> I delight . . . not to live in this restless, nervous, bustling, trivial Nineteenth Century, but stand or sit

thoughtfully while it goes by. What are men celebrating? They are all on a committee of arrangements, and hourly expect a speech from somebody. God is only the president of the day, and Webster is his orator.

In the end Thoreau restores the pastoral hope to its traditional location. He removes it from history, where it is manifestly unrealizable, and relocates it in literature, which is to say, in his own consciousness, in his craft, in *Walden*.

3

Hawthorne and Melville match the machine-in-the-garden motif to a darker view of life than Thoreau's. In their work, the design also conveys a sense of the widening gap between the facts and the ideals of American life, but the implications are more ominous. A nice illustration is Hawthorne's "Ethan Brand" (1850). This short story is particularly useful for an understanding of complex pastoralism and the experience that generates it, in spite of the fact that — or perhaps because — it exhibits only part of the motif. The pastoral ideal figures prominently in the tale, but the new technology does not. On the surface, at least, there is no indication that "Ethan Brand" embodies a significant response to the transformation of life associated with machine power. In fact, Hawthorne conceived of this fable as a variant of the Faust legend. Four years before writing it, in 1844, he had entered his *donnée* in his notebook. (It appears on the page immediately following the Sleepy Hollow episode.) "The search of an investigator for the Unpardonable Sin; — he at last finds it in his own heart and practice." And then, again, on the same page:

The Unpardonable Sin might consist in a want of love and reverence for the Human Soul; in consequence of which, the investigator pried into its dark depths, not with a hope or purpose of making it better, but from a cold philosophical curiosity, — content that it should be wicked in what ever kind or degree, and only desiring to study it out. Would not this, in other words, be the separation of the intellect from the heart? [14]

The tale is set on a lonely hillside in the Berkshires. Bartram, a doltish lime-burner, is sitting with his son at nightfall watching his kiln. They hear a roar of laughter — a chilling, mirthless laugh — and it frightens the child. We learn that the kiln formerly had belonged to one Ethan Brand, who, after gazing too long into the fire, had become possessed. A monomaniacal compulsion to seize an absolute truth had taken hold of him, and he had conducted a world-wide search for the Unpardonable Sin. He now appears, and with his bitter, self-mocking laugh announces the successful completion of his quest. He has found the worst of all sins in his own heart. An ironic circularity invests Brand's fate. Not only has he sought throughout the world for what was closest to himself, but, as it turns out, the Unpardonable Sin resides in the very principle for which he undertook the quest: the desire for knowledge as an end in itself.

At his father's behest, meanwhile, the boy has gone off to tell the "jolly fellows" in the tavern of Brand's return. Soon a whole shiftless regiment of drunks and derelicts appears on the mountain. Several of these worthies greet Brand and earnestly invite him to share their bottle, but he coldly rejects them. A series of minor incidents (a dog madly chasing its tail, as in *Faust*) underscores the tragic irony of Brand's life. After the crowd leaves and Bartram and his son have gone off to sleep, Brand sits alone beside the kiln reviewing his melancholy career. The compulsive quest has been a "success" in more than one sense. Not only has he found what he was seeking, but the search had been a means of education: Brand has become a world-renowned scholar. But he also recognizes that he has been transformed in the process. Once a simple man, capable of tenderness and pity, he is now an unfeeling, monomaniacal "fiend" who has lost his hold upon "the magnetic chain of humanity." With a cry of despair he throws himself into the fire. The next morning Bartram and his boy awake to a magnificent dawn. (A description of the landscape revealed by the early morning sun occupies two long paragraphs.) The mood of anxiety and gloom has been dispelled. As the boy suggests, the very sky and mountains seem to be rejoicing in Brand's disappearance. In the kiln they find his skeleton, converted into lime, though the shape of a human heart (was it made of marble?) remains visible between the ribs.

Taken by itself, the fable reveals no link between Brand's fate and Hawthorne's attitude toward industrialization. On the other hand, certain facts about the genesis of the story are suggestive. For many of the details, especially the setting and the portraits of several minor characters, Hawthorne drew upon notes he had made during a vacation journey in the Berkshires in 1838. At that time a small-scale industrial revolution was under way in the area. Textile production was increasing at a rapid rate, and near North Adams Hawthorne's stage-coach passed several new factories, ". . . the machinery whizzing, and girls looking out of the windows. . . ." Apparently fascinated by the sight, he took elaborate notes:

> There are several factories in different parts of North-Adams, along the banks of a stream, a wild highland rivulet, which, however, does vast work of a civilized

nature. It is strange to see such a rough and untamed stream as it looks to be, so tamed down to the purposes of man, and making cottons, woollens &c — sawing boards, marbles, and giving employment to so many men and girls; and there is a sort of picturesqueness in finding these factories, supremely artificial establishments, in the midst of such wild scenery. For now the stream will be flowing through a rude forest, with the trees erect and dark, as when the Indians fished there; and it brawls, and tumbles, and eddies, over its rock-strewn current. Perhaps there is a precipice hundreds of feet high, beside it, down which, by heavy rains or the melting of snows, great pine-trees have slid or tumbled headlong, and lie at the bottom or half-way down; while their brethren seem to be gazing at their fall from the summit, and anticipating a like fate. And taking a turn in the road, behold these factories and their range of boarding-houses, with the girls looking out of the window as aforesaid. And perhaps the wild scenery is all around the very site of the factory, and mingles its impression strangely with those opposite ones.[15]

The scene impresses Hawthorne with the sudden, violent character of change. In a single stroke a rude forest is being supplanted by "supremely artificial establishments," and all of nature seems poised, like the majestic pines, at the abyss. Here, as in the Berkshire notes generally, there is no place for a Sleepy Hollow, nor any time in which to realize the ideal middle landscape. Hawthorne traveled in the area for several days, and his observations, random sketches of places and people (including several "remarkable characters" who reappear in "Ethan Brand"), convey his sense of unreality and disorientation. One of the most revealing is this idea for a story:

A steam engine in a factory to be supposed to possess a malignant spirit; it catches one man's arm, and pulls it off; seizes another by the coat-tails, and almost grapples him bodily; — catches a girl by the hair, and scalps her; — and finally draws a man, and crushes him to death.

So far as we know, Hawthorne never developed this idea. And when he mined his Berkshire notes for "Ethan Brand," some ten years later, he passed over most of the economic and social data he had recorded. The factories do not appear in the story. Nor does it contain any explicit reference to the changing conditions of life — to industrialization. And yet there are more important ways in which the presence of the "machine" makes itself felt in the tale.[16]

A sense of loss, anxiety, and dislocation overshadows the world of "Ethan Brand." Initially, Hawthorne invests this feeling in his account of the lonesome lives of lime-burners and of the desolate, Gothic landscape. We are told that many of the kilns in this tract of country have been "long deserted," and that the wild flowers growing in the chinks make them look like "relics of antiquity." * A similar air of obsolescence clings to the villagers who climb the hillside to see Brand. All three selected for special mention are broken, unfulfilled men. One, formerly a successful doctor with a rare gift of healing, has become

---

* Ethan's kiln, later called a "furnace," is a link in a chain of virtually free association. It connects Hawthorne's response to the factories with his feelings about science and, more specifically, the revolutionary steam technology of the age. That the association transcends ordinary common-sense perception is indicated, for one thing, by the fact that the actual mills he saw near North Adams were operated by water — not steam — power. For another, the kilns in the tale serve as a curiously blended token of advancing technology and a mode of production rendered obsolete by that advance. The process of combining opposite meanings in a single image is akin to the "condensation" Freud discovers in dream symbols. The technological symbolism of "Ethan Brand" foreshadows the extraordinary fusion, in the whaling lore of *Moby-Dick*, of the most primitive hunting methods with advanced technical skills.

an alcoholic; another, "a wilted and smoke-dried man," representative of a vocation "once ubiquitous . . . now almost extinct," is the stage-agent; and a third, who also has known better days, is a lawyer who has come to be "but the fragment of a human being, a part of one foot having been chopped off by an axe, and an entire hand torn away by the devilish grip of a steam-engine." All of these people, along with a Wandering Jew and a forlorn old man searching for his lost daughter, are victims of change. Like the monomaniac Brand, whose "bleak and terrible loneliness" the young boy recognizes, each in his way is a maimed, alienated man. The center of the story is Brand's "cold philosophical curiosity," a disorder which finally results in a fatal "separation of the intellect from the heart."

But what is the cause of Brand's alienation? Of the two explanations Hawthorne provides, the first is mythic. According to local legend, Brand's intellectual obsession emanated from the fire. We learn that he "had been accustomed to evoke a fiend from the hot furnace of the lime-kiln," and that together they spent many nights evolving the idea of the quest. But the fiend, who may have been Satan himself, always retreated through the "iron door" of the furnace at the first glimmer of sunlight. Hawthorne manifestly borrows this image of fire, and the conventional contrast between fire and sun, from the literary tradition he knows best. The opposition between Satanic fire and life-giving sun, light of divine truth and righteousness, is a recurrent device in the work of Dante, Spenser, Milton, and Bunyan. The association of fire with technology, which goes back to Prometheus and Vulcan, also forms a part of the Christian myth. Before the fall, according to Milton, there was no need of technology. The garden contained only such tools "as art yet rude, / Guiltless of fire had formed." After the expulsion, however, Adam and Eve were grateful for the gift of fire.[17]

But if the contrast between fire and sun was conventional, it also was particularly meaningful to an age that called itself the "Age of Steam." William Blake had begun to exploit the new industrial associations of satanic fire in his *Milton* (1804–8), and later John Martin brought them into the pictures of hell he did for an illustrated edition of *Paradise Lost* (1827).* And it can be shown that Hawthorne uses the same images, words, and phrases to describe the fire in "Ethan Brand" that he uses elsewhere for direct reference to industrialization. A whole cluster of images surrounding the word "iron," including "fire," "smoke," "furnace," and "forge," serves to blend his feelings about his own age with the culture's dominant religious and literary tradition. As the Berkshire notes suggest, moreover, the introduction of industrial power in the American setting imparts a peculiar intensity to the dialectic of art versus nature. In this case the fire-sun antithesis provides the symbolic frame for the entire story: "Ethan Brand" begins at sundown and ends at dawn. During the long night the action centers upon the kiln, or "furnace," which replaces the sun as the origin of warmth, light, and (indirectly) sustenance. The fire

---

* This edition had a direct influence on the pastoral strain in American painting. Thomas Cole's *The Expulsion from the Garden of Eden* (1828) virtually was plagiarized from one of the illustrations. A sense of the opposition between industrial fire and pastoral sun makes itself felt in Cole's influential series, "The Course of Empire." The second canvas in the series, representing an explicitly pastoral state of social development, portrays the era of harmony and tranquillity just before the decline of civilization begins.

imagery aligns Ethan's fate simultaneously with the Christian doctrine of sin and with the scientific-industrial revolution. Only when the fire has been extinguished does Hawthorne introduce a vision of the garden.[18]

That fire is a surrogate for the "machine" in this variant of the Sleepy Hollow motif becomes more apparent as the story nears its climax. In describing Ethan's reflections while he sits alone beside the kiln, Hawthorne supplies a second (historical) explanation for his fall. This version makes the protagonist an embodiment of a changing America. It recalls the view of history put forward by Jefferson when, speaking of the War of 1812, he had awarded the British enemy the "consolation of Satan" for helping to transform a peaceable, agricultural nation into a military and manufacturing one. Now Ethan Brand (an archetypal Yankee name joined to an image of burning and infamy) sees it all clearly. Once, when "the dark forest had whispered to him," he had been "a simple and loving man." He had felt tenderness and sympathy for mankind. But then — here is the point, presumably, when the fiend had emerged from the fire — he had begun, with the highest humanitarian motives, "to contemplate those ideas which afterwards became the inspiration of his life." In this telling Hawthorne abandons the language of myth. Alluding to neither fiend nor fire, he now attributes Ethan's obsession to a "vast intellectual development, which . . . disturbed the counterpoise between his mind and heart." With the weakening of his moral nature he had become "a cold observer, looking on mankind as the subject of his experiment." In other words, Ethan is at once an agent and a victim of scientific empiricism or "mechanism." (The etiology of his obsession resembles Carlyle's account of the characteristic psychic imbalance of the "Age of Machinery.") The Unpardonable Sin is the great sin of the Enlightenment — the idea of knowledge as an end in itself. Now he recognizes the destructiveness of the idea. Lonely, desperate, alienated from nature and mankind, he plunges to his death in the kiln.

During the night the fire goes out, and this is the scene when Bartram and his son arise at daybreak:

> The early sunshine was already pouring its gold upon the mountain-tops, and though the valleys were still in shadow, they smiled cheerfully in the promise of the bright day that was hastening onward. The village, completely shut in by hills, which swelled away gently about it, looked as if it had rested peacefully in the hollow of the great hand of Providence. Every dwelling was distinctly visible; the little spires of the two churches pointed upwards, and caught a fore-glimmering of brightness from the sun-gilt skies upon their gilded weather-cocks. The tavern was astir, and the figure of the old, smoke-dried stage-agent, cigar in mouth, was seen beneath the stoop. Old Graylock was glorified with a golden cloud upon his head. Scattered likewise over the breasts of the surrounding mountains, there were heaps of hoary mist, in fantastic shapes, some of them far down into the valley, others high up towards the summits, and still others, of the same family of mist or cloud, hovering in the gold radiance of the upper atmosphere. Stepping from one to another of the clouds that rested on the hills, and thence to the loftier brotherhood that sailed in air, it seemed almost as if a mortal man might thus ascend into the heavenly regions. Earth was so mingled with sky that it was a day-dream to look at it.
>
> To supply that charm of the familiar and homely, which Nature so readily adopts into a scene like this, the stage-coach was rattling down the mountain-road,

and the driver sounded his horn, while Echo caught up
the notes, and intertwined them into a rich and varied
and elaborate harmony, of which the original performer
could lay claim to little share. The great hills played a
concert among themselves, each contributing a strain of
airy sweetness.

The immediate effect of this radiant prospect is to
heighten the sense of evil surrounding Ethan Brand and
his quest. It all too abundantly represents the American
Eden from which he has fallen. Hawthorne plays every
detail against the preceding Walpurgisnacht. Instead of
fire and smoke, there is perfect sunlit visibility; instead
of anxiety and gloom, the countryside is invested with order, peace, and permanence, and the image of a stepladder
into the heavens projects the harmony between man and
nature beyond nature. No traces of the "machine" remain.
Having referred to the imminent disappearance of the
stagecoach, Hawthorne now restores that picturesque vehicle to its proper place in the New England landscape. Like
the coming of spring at the end of *Walden,* this eighteenth-century tableau figures a "realization of the Golden
Age." With the elimination of Ethan Brand and everything that he represents, Hawthorne seems to be saying,
the pastoral ideal has been realized.

But it is apparent that the perceptive reader is not expected to take this grand picture of the morning sun
"pouring its gold" upon the land entirely at face value.
For a moment, to be sure, Hawthorne may seem to be inviting the obvious response. But the florid, unabashedly
trite language should make us wary. What Melville said
of other aspects of Hawthorne's stories may be applied to
his narrative voice: it seems to have been "directly calculated to deceive — egregiously deceive — the superficial

skimmer of pages." One has only to sample the fiction in contemporary magazines, especially the ladies books for which he often wrote, to see how perfectly Hawthorne has caught the sickly sweet, credulous tone of sentimental pastoralism. In fact, he has assembled a splendid exhibit of bucolic clichés. They are all here, the gold laid on thick, the hills swelling gently, the heaps of hoary mist on breasts of mountains, the charms that "Nature . . . adopts into a scene like this" — "it was a day-dream to look at it" — and, dominating the whole glorious panorama, the rich, varied, elaborate harmony of Echo.* Removed from its dramatic context this set piece reads like parody: "The great hills played a concert among themselves, each contributing a strain of airy sweetness." [19]

But what, then, is the relation between the subtle meretriciousness of the pastoral moment and the suicide of Ethan Brand? That Hawthorne means the juxtaposition to be ironic is evident. On the basis of the tale alone,

* It is fitting that Hawthorne by-passes the relatively fresh and vigorous rhetorical mode of Thoreau, Emerson, and Wordsworth. The moribund idiom he uses may be traced from American popular culture back, by way of writers like Irving, to Goldsmith, Gray, and, for that matter, as far back as Thomson's "Spring" (1728) — the poem admired by young Jefferson. Consider, for example, this passage:

> Harmonious Nature too look'd smiling on.
> Clear shone the skies, cool'd with eternal gales,
> And balmy spirit all. The youthful sun
> Shot his best rays, and still the gracious clouds
> Dropp'd fatness down; as o'er the swelling mead
> The herds and flocks, commixing, play'd secure.
>
> .   .   .   .
>
> For music held the whole in perfect peace:
> Soft sigh'd the flute; the tender voice was heard,
> Warbling the varied heart; the woodlands round
> Applied their choir; and winds and waters flow'd
> In consonance.

however, it is difficult to know what the irony signifies. Perhaps because the author himself was uncertain, or because he wanted it to work both ways, the ending is unusually muted. For that matter the entire story gives an impression of constricted thought. Of all Hawthorne's well-known stories, "Ethan Brand" is the least polished: the action is spasmodic; the texture jagged; the irony severely understated; the conclusion elliptical. In places one can see where passages from the notebooks have been imperfectly joined. Yet the story is crammed with unrealized power — fairly bursting with it, like the synopsis of a Wagnerian libretto. Perhaps this is because it was written at a turning point in Hawthorne's career, when he was about to shift from the short story to the novel-length romance. In fact, there are good reasons for thinking that it was intended as a longer work. (He gave it the subtitle, "A Chapter from an Abortive Romance.") At any rate, the constrained, fettered manner reflects Hawthorne's state of mind at the time. In 1848, apparently referring to "Ethan Brand," he told a New York editor: "At last, by main strength, I have wrenched and torn an idea out of my miserable brain; or rather, the fragment of an idea, like a tooth ill-drawn, and leaving the roots to torture me." [20]

The meaning of Hawthorne's complex version of pastoral is to be found among the torn roots of his ideas. In writing "Ethan Brand," he had left behind the manifestly topical content of his 1838 Berkshire notes, but he had transferred certain ideas and emotions once attached to that material to other themes and images. Although the striking sight of factories in the wilderness does not appear in the tale, Hawthorne's feelings about it do. Ethan is destroyed by the fires of change associated with factories, and nothing confirms this fact as forcibly as the meretricious idyllic vision that follows his death. Having observed the illusoriness of the middle landscape ideal in reality (the wilderness is to be supplanted by factories, not gardens), Hawthorne now accomplishes its restoration in fiction. Needless to say, the restoration is ironic. As the hollow rhetoric indicates, this stock eighteenth-century tableau serves as an oblique comment upon the fate of the Yankee hero. It says that the dream of pastoral harmony will be easy to realize as soon as the Faustian drive of mankind — the Brand element — has been extirpated. Nor is the perfervid idiom of sentimental pastoralism the only clue to the irony. It is reinforced by such details as the histrionic stagecoach, a vehicle supplied, absurdly enough, by "Nature." To fulfill the pastoral hope, in other words, nothing less is required than a reversal of history. "Ethan Brand" conveys Hawthorne's inchoate sense of the doom awaiting the self-contained village culture, not the institutions alone, but the whole quasi-religious ideology that rests, finally, upon the hope that Americans will subordinate their burning desire for knowledge, wealth, and power to the pursuit of rural happiness. Hence the pretty picture that accompanies Ethan's death. "The village, completely shut in by hills, which swelled away gently about it, looked as if it had rested peacefully in the hollow of the great hand of Providence."

"In a week or so," Melville wrote to Hawthorne in June 1851, "I go to New York, to bury myself in a third-story room, and work and slave on my 'Whale' while it is driving through the press." And then, in the next paragraph: "By the way, in the last 'Dollar Magazine' I read 'The Unpardonable Sin.' He was a sad fellow, that Ethan Brand. I have no doubt you are by this time responsible for many a shake and tremor of the tribe of 'general readers.'" [21]

Melville's response to Hawthorne's story deserves attention because, for one thing, Ethan's fire reappears in the brick and mortar fireplaces, or try-works, aboard the *Pequod*. Speaking of these "furnaces," the industrial center of the whaling enterprise and a distinguishing feature of American whalers, Ishmael says, "It is as if from the open field a brick-kiln were transported to her planks." In composing "The Try-Works," a decisive chapter in Ishmael's repudiation of Ahab's quest, Melville incorporates the opposition between fire and sun — and with it a sense of the transformation of American society — into the pattern of contradiction that controls the meanings of *Moby-Dick*. His reaction to Hawthorne's story is typical of his high-keyed, impressionable state of mind during the spring and summer of 1851. In this remarkable letter he is manifestly assimilating ideas he finds in "Ethan Brand" into the complex pastoral design in *Moby-Dick*. The letter continues:

> It is a frightful poetical creed that the cultivation of the brain eats out the heart. But it's my *prose* opinion that in most cases, in those men who have fine brains and work them well, the heart extends down to hams. And though you smoke them with the fire of tribulation, yet, like veritable hams, the head only gives the richer and the better flavor. I stand for the heart. To the dogs with the head! I had rather be a fool with a heart, than Jupiter Olympus with his head. The reason the mass of men fear God, and *at bottom dislike* Him, is because they rather distrust His heart, and fancy Him all brain like a watch.

In the next sentence Melville changes the subject, and then he changes it again and again in a stunning spiral of associations. A comment upon Hawthorne's growing fame leads to a summary of his own intellectual development, and that is followed by a tribute to the wisdom of Solomon (also to reappear in "The Try-Works"), and the letter concludes with a superbly cogent statement of his complicated attitude toward the transcendental idea of nature. In a half-conscious way, Melville apparently is following the obscure logic at work in Hawthorne's tragic pastoral. At least this much is clear: an awareness of the opposition between head and heart, fire and sun, leads him to reconsider the claims of contemporary pastoralism:

> — In reading some of Goethe's sayings, so worshipped by his votaries, I came across this, *"Live in the all."* That is to say, your separate identity is but a wretched one, — good; but get out of yourself, spread and expand yourself, and bring to yourself the tinglings of life that are felt in the flowers and the woods, that are felt in the planets Saturn and Venus, and the Fixed Stars. What nonsense! Here is a fellow with a raging toothache. "My dear boy," Goethe says to him, "you are sorely afflicted with that tooth; but you must *live in the all*, and then you will be happy!" As with all great genius, there is an immense deal of flummery in Goethe, and in proportion to my own contact with him, a monstrous deal of it in me.
>
> H. Melville.
>
> P.S. "Amen!" saith Hawthorne.
>
> N.B. This "all" feeling, though, there is some truth in. You must often have felt it, lying on the grass on a warm summer's day. Your legs seem to send out shoots into the earth. Your hair feels like leaves upon your head. This is the *all* feeling. But what plays the mischief with the truth is that men will insist upon the universal application of a temporary feeling or opinion.

This incandescent letter lights up the relation between *Moby-Dick* and the spurious pastoralism of the age. Like Hawthorne, Melville is in effect obeying Emerson's injunction to wise men to "pierce this rotten diction and fasten words again to visible things." * But Melville is bolder and more subversive than Hawthorne. While Hawthorne hints guardedly at the false character of the essentially moribund, Augustan pastoralism of the dominant culture, Melville's witty attack embraces the flummery of the romantic *avant-garde* as well — including, to a degree, himself. Indeed, what enables him to write this astonishing letter is unsparing self-knowledge; he is exposing the basis of pastoral fantasying in himself, and his insight illuminates *Moby-Dick*. The final sentence alone, which contains the gist of what Ishmael learns aboard the *Pequod*, neatly circumscribes the romantic pastoralism, always verging upon pantheism, of Goethe, Wordsworth, and Emerson. The extravagant claims of that doctrine, Melville is saying, stem from a tendency to confuse a transitory state of mind — the "all" feeling — with the

* Emerson's remark occurs in a seminal passage in *Nature* (1836; ch. IV) which looks forward to George Orwell's demonstration of the reciprocal relation between corrupt politics and corrupt language. Emerson argues that the "corruption of man is followed by the corruption of language," and he goes on to indicate why a particular idiom — a pastoral idiom, we might say — can be adequate to one age and corrupt in the next. "When simplicity of character . . . is broken up by the prevalence of secondary desires, — the desire of riches, of pleasure, of power, and of praise, — and duplicity and falsehood take place of simplicity and truth, the power over nature as an interpreter of the will is in a degree lost; new imagery ceases to be created, and old words are perverted to stand for things which are not; a paper currency is employed, when there is no bullion in the vaults. In due time the fraud is manifest, and words lose all power to stimulate the understanding or the affections. Hundreds of writers may be found in every long-civilized nation who for a short time believe and make others believe that they see and utter truths, who do not of themselves clothe one thought in its natural garment, but who feed unconsciously on the language created by the primary writers of the country, those, namely, who hold primarily on nature.

But wise men pierce this rotten diction and fasten words again to visible things. . . ."

universal condition of things. He does not deny the significance of that religious emotion as a gauge of man's inner needs; what he does attack is the mistake of projecting it upon the universe. The letter is a treatise in small against excessive trust in what Freud calls the "oceanic feeling." With one clean, surgical stroke Melville severs the nineteenth century's burgeoning, transcendental metaphysic from its psychic roots.[22]

Melville's insight was, of course, not the product of sudden inspiration. Granted that his encounter with Hawthorne and with Hawthorne's writing in 1850 had had a catalytic effect, the truth is that he had been working toward this moment of creative equilibrium ever since his return from the Pacific in 1844. Until that year, he writes in a now familiar passage from the same letter, "I had no development at all. From my twenty-fifth year I date my life. Three weeks have scarcely passed, at any time between then and now, that I have not unfolded within myself. But I feel that I am now come to the inmost leaf of the bulb, and that shortly the flower must fall to the mould." In 1846 he had begun his literary career with *Typee* and (though the fact is obscured by the genre of the travel romance) with an expression of the pastoral impulse. At the outset the narrator, Tommo, has been at sea for six months without a sight of land: "Oh! for a refreshing glimpse of one blade of grass — for a snuff at the fragrance of a handful of the loamy earth!" he exclaims; "Is there no green thing to be seen?" Ishmael, the narrator, asks a similar question at the beginning of *Moby-Dick*. He is explaining his decision to embark upon an ocean voyage. At the time he had been in a bleak, depressed mood, and going to sea, he admits, had been his substitute for suicide. He thinks that most people (if they but knew it) cherish similar feelings about the ocean.

On a dreamy Sabbath afternoon, he says, you will see clusters of men standing all around Manhattan, fixed like sentinels in ocean reveries. Yet there is something puzzling about their behavior. Why are they attracted to the water? After all, they are landsmen "of week days pent up in lath and plaster — tied to counters, nailed to benches, clinched to desks. How then is this? Are the green fields gone?"

Melville's development between *Typee* and *Moby-Dick* may be described as an unfolding of the ideas and emotions packed into this question. It is Ishmael who will enact the final unfolding. Even before going to sea, it appears, he had had an intimation of the kinship between his feeling for the ocean and a sense of green fields gone; but it had been a mere intimation, and only his part in Ahab's quest enables him, finally, to grasp its meaning. Thus, at the beginning of *Moby-Dick,* Melville restates the lament with which he had begun *Typee.* Although Tommo is aboard ship and Ishmael on shore, each narrator is introduced at the moment his yearning for greenness impels him to forsake an organized community. In each case, moreover, the lament is put in the form of a question, so that even as the narrator surrenders to the escapist impulse he is uncertain about its validity. A hint of the complex pastoral motive in *Moby-Dick* is implicit in Tommo's situation on the first page of Melville's first book. He is an alienated American seafarer aboard a whaling ship; the ship, an extension of Western civilization, is moving across an inscrutable Pacific wilderness, and the hero's dream of felicity is made manifest in images of an evanescent green land.

Melville's distance from the soft pastoralism of the age is implicit in this spatial scheme: it excludes the middle landscape as a token of a realizable ideal. In *Typee* the hero's retreat carries him directly into primal nature. With a friend, Tommo jumps ship in the Marquesas and, after certain trials, finds himself living among the primitive Typees. In spite of a reputation for savagery, these handsome, easy-going people strike him as charming. They charm him in the same way that the Virginia Indians had charmed Robert Beverley: by their capacity for joy. They enjoy life in a Gauguin world of eternal summer and natural abundance. And so for awhile does Tommo. There are no calendars or timetables or clocks in the valley, and the hours (as at Walden or aboard Huck Finn's raft) trip along "as gaily as the laughing couples down a country dance." At every turn Tommo is reminded of other items of American life whose absence means pleasure. There is no money, no business, no poverty — "none of those thousand sources of irritation that the ingenuity of civilized man has created to mar his own felicity." And most delightful, there is scarcely any work. "The penalty of the Fall," says Tommo, "presses very lightly upon the valley of Typee. . . ." This "paradisiacal abode" is not to be confused with the cultivated, bucolic domain of the Jeffersonian myth. The garden Tommo has discovered is a primitive, prelapsarian garden: no property, no letters, no work — "no occupation; all men idle, all." It very nearly fulfills Gonzalo's requirements for a model plantation:

> All things in common Nature should produce
> Without sweat or endeavor: treason, felony,
> Sword, pike, knife, gun, or need of any engine,
> Would I not have; but Nature should bring forth,
> Of it own kind, all foison, all abundance,
> To feed my innocent people.

The extravagance of Tommo's fantasies, their resemblance to those of an Elizabethan voyager, is a gauge of Melville's prescient historical imagination. At a time when the Amer-

ican Republic remains, quite literally, a half vacant land
of green fields, his hero sounds less like a hopeful, Emersonian Young American than like a fugitive from a highly
developed, not to say over-developed, civilization. The
United States, Melville writes in his 1850 essay on Hawthorne, is "rapidly preparing for that political supremacy
among the nations, which prophetically awaits . . . [it]
at the close of the present century. . . ." [23]

In *Typee*, as in *The Tempest*, the movement toward
nature is checked by Caliban — by a Melvillean counterpart, that is, to Shakespeare's "thing of darkness." The
turning point is Tommo's discovery that the Typees are
in fact cannibals. In a series of quiet but sinister episodes
Melville leads his hero to the edge of primitive terror.
The natives become increasingly possessive. They try to
tattoo his face. They guard him like a captive. As his
suspicions rise, the utter mindlessness of his existence, the
absence of anyone with whom he can talk (his shipmate
having gone off and not returned), intensifies his desperation. Pleasant as they are, his relations with the natives
are merely external. He does not, cannot, know their
minds. He is alone. Besides, upon arriving in the valley
he had had a mysterious ailment in his leg, and now, as
fear grows upon him, the symptoms recur with more violence than ever. In view of the overtones with which Melville later surrounds Ahab's loss of a leg, Tommo's illness
strongly suggests a conflict involving sexual guilt, repression, and impotence. It "nearly unmanned me," he says
of the pain in his leg. In this primitive garden he has
enjoyed an erotic freedom unthinkable in nineteenth-
century America; the psychosomatic consequence of the
retreat into nature is to accentuate his need for the healing power of art — in this case, medical art. *Typee* is an

ingenuous book that falls somewhere between fiction and autobiographical confession, and Melville may be observed in the process of discovering an equivalence between the exterior and the psychic landscapes. At the climax Tommo's manhood is threatened by "cannibalism" in both terrains: the ferocity of flesh-eating natives is matched by a nameless aggressive force within himself. In the end he commits an act of violence to escape from the "freedom" of nature.[24]

The question Tommo had posed in the beginning — "Is there no green thing to be seen?" — is not fully answered in *Typee,* nor is it answered with anything like finality in the other oceanic pastorals that Melville writes before *Moby-Dick.* But its implications are enriched, and by 1850 it has come to mean something like this: can men hope to find a green field, a place or condition in which they are not in danger of being nailed to workbenches or cooked in fleshpots? Is it possible to mediate the claims of our collective, institutional life and the claims of nature? Here greenness is a token of earthly felicity akin to the pastoral hope. But its meaning cannot be understood apart from the rest of the Melvillean world. By the time he begins work on "The Whale," Melville has delineated a symbolic setting adequate to the complexity and clarity of his thought. It is divided, provisionally at least, into three realms: (1) a ship, mobile replica of a technically advanced, complex society; (2) an idyllic domain, a lovely green land that figures a simple, harmonious accommodation to the conditions of nature, and (3) a hideous, menacing wilderness, habitat of cannibals and sharks located beyond (or hidden beneath the surface of) the bland green pastures. To answer his question ("Are the green fields gone?") Ishmael must reconcile the apparent contradiction between these two states of nature. Like Shakespeare's American fable, *Moby-Dick* is an exploration of the nature of nature.

In this fictive world the ship occupies a place like that of "Concord" in *Walden*. Even in *Typee*, where Melville does not develop the device of the ship as a small-scale social mechanism, shipboard life is joyless, routinized, dull. Tommo's longing for a sight of green grass is a reaction to the "privations and hardships" aboard the whaler. Later, Melville self-consciously exploits the trope of the ship as a social microcosm. He gives *White-Jacket* (1850) the subtitle, "The World in a Man-of-War," and toward the end he introduces an image of collective power that was to have a decisive function in *Moby-Dick*: "The whole body of this discipline," says the narrator, summing up his impressions of life aboard an American frigate, "is emphatically a system of cruel cogs and wheels, systematically grinding up in one common hopper all that might minister to the moral well-being of the crew." In *Moby-Dick* he uses the image of an implacable machine to express Ahab's pride in dominating the *Pequod*'s crew. In this he reflects the attitude of Carlyle and the romantics. (During the summer of 1850 Melville apparently read Carlyle.) The creed of the Age of Machinery, Carlyle had said in "Signs of the Times," is Fatalism: by insisting upon the force of circumstances, men argue away all force from themselves, until they "stand lashed together, uniform in dress and movement, like the rowers of some boundless galley." By the time Ahab sights the white whale, the *Pequod* has become such a fated ship. It is Ahab's ship in every possible sense. He helps to create the "circumstances" which in turn form the crew into an instrument of his metaphysical intent. In the try-works episode Ishmael realizes that the whaler has been transformed into a "fire-ship," the "material counterpart of her monomaniac commander's soul." And by the second day of the chase the crew has become a pliant, disciplined, committed, totalitarian unit. "They were one man, not thirty . . . all varieties . . . welded into oneness, and . . . all directed to that fatal goal which Ahab their one lord and keel did point to." [25]

All are directed toward Ahab's goal, that is, except Ishmael. In the end he alone is saved, a fact that comports with his success in establishing a position independent of Ahab and the fiery quest. The significance that Melville attaches to Ishmael's survival is indicated by the line from Job he takes as the motto of the Epilogue: "And I only am escaped alone to tell thee." In other words Ishmael's relation to us, the readers of *Moby-Dick*, is like that of Job's messengers to Job. The calamity he recounts is a portent of further trials to come: we too may expect our integrity and faith to be tested. Ishmael's survival — the contrast between his attitude and Ahab's — is the root of the matter. But the contrast develops only in the course of the voyage: earlier, Ishmael tells us, his views had had a good deal in common with Ahab's. His "splintered heart and maddened hand" also had been "turned against the wolfish world," and his chief motive for undertaking the voyage, moreover, had been "the overwhelming idea of the great whale himself." Long before he encountered Ahab, that "portentous and mysterious monster" had roused all his curiosity; he had been swayed to the purpose by wild conceits, "endless processions of the whale, and, mid most of them all, one grand hooded phantom, like a snow hill in the air." And when the crew, roused by the eloquence of Ahab's quarter-deck speech, swears death to Moby Dick, Ishmael's shouts go up with the rest; his oath is "welded" with theirs. Before the end, however, he changes his mind. He discovers that the "green fields," although they do not signify the literal possibility of a pastoral life, are not gone. Ishmael's rediscovery of greenness is of a piece with his repudiation of the "horrible oath": the action that leads to the *Pequod*'s doom results in his "salvation." [26]

As the movement toward tragedy gains momentum, all meanings are polarized, so that we are left, finally, with the irreconcilable views of Ahab and Ishmael. Their contrasting ideas of man's relation to nature correspond to the conflicting attitudes which Melville had expressed, along with his reaction to "Ethan Brand," in the letter to Hawthorne. His initial scorn of romantic pastoralism as a ludicrously feeble defense against despair is embodied in Ahab. To a man as angry as Ahab — a desperate, incomplete, mortally wounded ("unmanned") man, a man who suffers the pain of unbelief — the "all" principle hardly provides consolation. But if the *Pequod*'s captain represents the side of Melville that says "what nonsense!" Ishmael lends expression to the second thoughts of his postscript: "This 'all' feeling, though, there is some truth in. You must often have felt it, lying on the grass on a warm summer's day."

The course of Ishmael's inner development, which is diametrically opposed to Ahab's, leads to the discovery of that truth. Melville underscores the divergence by introducing the two characters in contrasting situations. When the narrative begins, Ishmael's sense of alienation from greenness is most acute: he is a desperate man. On the other hand, when Ahab announces his purpose he is at the height of his powers, master of the ship and of the crew, seeming master of himself, a Titan prepared to defy the absolute. "Talk not to me of blasphemy, man," he replies to Starbuck. "I'd strike the sun if it insulted me." But as Ahab's heart hardens against the world, Ishmael undergoes a "melting within." To point the contrast Melville withholds Ahab's version of Ishmael's opening lament until the end. Only in "The Symphony," the elegiac interlude on the eve of the chase, does Ahab fully grasp the relation between his seafaring life and his monomania. Meanwhile, Ishmael has freed himself of despair; he has located the "green fields" — the germ of truth in the "all" feeling.

The sea change that Ishmael suffers, played against the opposite change in Ahab, is the narrative key to the pastoral design in *Moby-Dick*. Telling his story in retrospect, Melville's narrator speaks as one who has avoided the trap of sentimental pastoralism. Before the voyage he had been a typical young romantic with an itch for remote places and a tendency to pantheistical revery. He tells us that he had embarked in search of "an asylum . . . [from] the carking cares of earth, and seeking sentiment in tar and blubber." Instead of an asylum, he had found a hideous wilderness. His violent recoil from society, like Tommo's, had been checked only at the brink of primitive mindlessness. Unable to locate a haven at sea or on land, Ishmael had been impelled toward a new definition of felicity or greenness. It is in dramatizing this process that Melville introduces a version of the design Hawthorne had used in his Sleepy Hollow notes. At three decisive points he incorporates the device into the larger, tragic pattern of contradiction.[27]

Melville's first significant use of the design occurs in the juxtaposition of two important episodes in chapters 35 through 37. Ishmael's account of his first masthead watch, when he discovered the dangers of romantic pastoralism, is followed immediately by Ahab's ringing announcement of the quest for the white whale, and by the soliloquy in which he exults in his power over the crew. In this ironic counterpoint of scenes Melville relies upon two images: the ocean's surface pictured as an endless series of "watery pastures" and an invincible locomotive conquering the landscape.

In "The Mast-Head" Ishmael describes his momentary surrender to the "all" feeling. It was his first watch aloft. Standing high above the deck on a serene, tropical day, one is likely to become lost, he says, in the infinite series of the sea. A "sublime uneventfulness invests you" as you gaze out across the undulating "watery pastures." Lulled into "an opium-like listlessness of vacant, unconscious revery . . . by the blending cadence of waves with thoughts," you soon lose your identity. What Melville is describing, of course, is a consummate Emersonian experience of nature. Natural facts are blended with Ishmael's thoughts by the rhythm of the sea, and so he merges with the all. "In this enchanted mood," he says, "thy spirit ebbs away to whence it came; becomes diffused through time and space. . . ." Or, as Emerson had said of a similar experience, "I am nothing; I see all; the currents of the Universal Being circulate through me; I am part or parcel of God." In Melville's case, however, the interlude does not lead to an affirmation of a transcendent meaning. On the contrary, "The Mast-Head" is designed to show that such fantasies may end in horror.[28]

Melville's theme, which he had introduced at the very beginning of the book — shortly after Ishmael's query, "are the green fields gone?" — is that the romantic attitude toward external nature is finally narcissistic. This point, Ishmael then had suggested, is more clearly revealed by our responses to "watery pastures" than to ordinary landscapes. Continuing his attempt to account for the fascination exercised upon mankind by the ocean, Ishmael had proposed the theory that water is the telltale element in our relation with visible nature. To clinch the argument he had introduced the example of a painter of pastoral landscapes:

> He desires to paint you the dreamiest, shadiest, quietest, most enchanting bit of romantic landscape. . . . What is the chief element he employs? There stand his trees, each with a hollow trunk, as if a hermit and a crucifix were within; and here sleeps his meadow, and there sleep his cattle; and up from yonder cottage goes a sleepy smoke. Deep into distant woodlands winds a mazy way, reaching to overlapping spurs of mountains bathed in their hill-side blue. But though the picture lies thus tranced, and though this pine-tree shakes down its sighs like leaves upon this shepherd's head, yet all were vain, unless the shepherd's eye were fixed upon the magic stream before him.

The shepherd's eye should be fixed upon the "magic stream," Ishmael is saying, because that image typifies our endless fascination with visible nature. He singles out water for its "magical" properties: at times transparent, at others a mirror, water bemuses us with the possibility of penetrating the surface of nature, yet it flatters and disturbs us by casting back our own image. What do we actually see — the object or ourselves? A shepherd gazing at the water thus epitomizes the exasperating, yet perhaps unavoidable, mingling of the "I" with the object in the act of perception. All of which is illustrated, Ishmael says, by the story of Narcissus, who, because he could not grasp the tormenting image he saw in the fountain, plunged into it and was drowned. What makes water the telltale element in landscape is that it so clearly elicits the narcissistic response. The image that Narcissus saw in the fountain, Ishmael concludes, "we ourselves see in all rivers and oceans. It is the image of the ungraspable phantom of life; and this is the key to it all."

In "The Mast-Head" Ishmael describes how dangerously close he had come to repeating the fatal plunge of Narcissus. He had been one of those "romantic, melancholy, and absent-minded young men" — an emulator of Childe Harold — who are useless as look-outs. While perched high above the sea, taking "the mystic ocean at his feet for the visible image of that deep, blue, bottomless soul, pervading mankind and nature," Ishmael had forgotten the whales and other monsters swimming beneath him. As the closing sentences of the chapter reveal, however, the voyage of the *Pequod* had taught him the perilousness of such transcendental moments:

> But while this sleep, this dream is on ye, move your foot or hand an inch; slip your hold at all; and your identity comes back in horror. Over Descartian vortices you hover. And perhaps, at mid-day, in the fairest weather, with one half-throttled shriek you drop through that transparent air into the summer sea, no more to rise for ever. Heed it well, ye Pantheists!

The horrifying idea of a fall from the heights of pastoral revery into the undersea vortex of material reality is the counterpart, in this variant of the motif, to the railroad's sudden incursion in Sleepy Hollow. "Heed it well, ye Pantheists!" is Ahab's cue; it provides the transition to "The Quarter-Deck" and the long-delayed announcement of Ahab's purpose. Assembling the entire ship's company one late afternoon, Ahab shows a gold doubloon to the men. He now displays a curious affinity with the "machinery" of motivation in this acquisitive society. While waiting for Starbuck to get a hammer, Ahab hums to himself, "producing a sound so strangely muffled and inarticulate that it seemed the mechanical humming of the wheels of his vitality in him." Then, announcing that the doubloon is for the man who first raises the white whale, Ahab nails the coin to the mast. Only Starbuck objects: ". . . I came here to hunt whales, not my commander's vengeance. How many barrels will thy vengeance yield . . .?" In putting down the recalcitrant mate, Ahab makes the electrifying speech that so vividly expresses his Faustian compulsion to impose his will upon the cosmos. In Melville's hero the thrust of Western man for ultimate knowledge and power is sinewed with hatred. Ignorance of the absolute, for Ahab, is a humiliation; he equates not knowing with being senselessly dismembered or imprisoned. (Here, in rapid succession, he speaks of the whale as a malicious dismemberer, a prison wall, a mask concealing the ultimate secret.) Nothing could be further from the quiet tenderness associated with the pastoral feeling for nature; Ahab's vengeful drive is directed not toward but through the visible universe. In one unforgettable metaphor he sums up the reckless, violent irreverence implicit in the "progress" of "advanced" civilization. "All visible objects, man," he cries, "are but as pasteboard masks. . . . If man will strike, strike through the mask!" Then, his eloquence having roused the fervor of the crew, and even the bewitched, tacit acquiescence of Starbuck, Ahab retires to his cabin.

The chapter that follows, "Sunset," is a brief poetic

soliloquy on power. Ahab is sitting alone beside the stern windows, and as he gazes out the *Pequod*'s wake is transformed, in his mind's eye, into the "track" of a nameless, inexorable force cutting through the sea. "I leave a white and turbid wake; pale waters, paler cheeks, where'er I sail. The envious billows sidelong swell to whelm my track; let them; but first I pass." No visible objects, no physical reality, can deter him. But the sun is setting, and Ahab's dream of power is a reminder of his incapacity for pleasure.

> Oh! time was, when as the sunrise nobly spurred me, so the sunset soothed. No more. This lovely light, it lights not me; all loveliness is anguish to me, since I can ne'er enjoy. Gifted with the high perception, I lack the low, enjoying power; damned, most subtly and most malignantly! damned in the midst of Paradise!

Only the thought of dominion eases the pain, and he calls up the sweet memory of Starbuck's capitulation: "I thought to find one stubborn, at the least; but my one cogged circle fits into all their various wheels, and they revolve." Then, again, he sees himself as igniting a fiery explosion, the crew "like so many ant-hills of powder, they all stand before me; and I their match." The thought of being consumed in the act provokes a reassertion of his will; he vows to "dismember my dismemberer," and as fanatic determination rises in him, he defies the gods to come out of their hiding places, taunts them in the rhetoric of the technological sublime:

> Come, Ahab's compliments to ye; come and see if ye can swerve me. Swerve me? ye cannot swerve me, else ye swerve yourselves! man has ye there. Swerve me? The path to my fixed purpose is laid with iron rails, whereon my soul is grooved to run. Over unsounded gorges,

through the rifled hearts of mountains, under torrents'
beds, unerringly I rush! Naught's an obstacle, naught's
an angle to the iron way!

The effect of these counterpointed episodes — Ishmael's
daydream and the sudden explosion of Ahab's muffled
aggression — is to sharpen the border line between illusion and reality in Melville's fictive world. On the mast-
head, Ishmael, forgetting about the undersea wilderness
of sharks and whales, had lost his identity in the "upper
air" of pure consciousness. But Ahab's charismatic speech
had brought him back to material reality. "For one," he
admits, "I gave myself up to the abandonment of the time
and the place; but while yet a-rush to encounter the whale,
could see naught in that brute but the deadliest ill." Feeling a "vague, nameless horror" concerning the whale, and
conscious of his responsibility to make credible the fiery
hunt (he is, after all, the narrator of this incredible story),
Ishmael turns his attention to the hard facts. He under-
takes a systematic study of the entire whaling enterprise.
In *Moby-Dick* the governing metaphor of reality is the
violent encounter with the whale.[29]

In a whaling world, Ishmael discovers, man's primary
relation to nature is technological. No detail figures this
truth more clearly than the "magical, sometimes horrible
whale-line" he discusses in "The Line" (Chapter 60). This
relatively insignificant passage — insignificant so far as the
action is concerned — illustrates the astonishing range of
insight released by Melville's whaling trope. By describing the line that follows the harpoon, he discloses the
elemental aspect of physical dependence, plunder, and exploit that underlies the deceptively mild, abstract quality
of life in our technical civilization. Here the simple

Manila rope is made to seem an archetype of the physical bond between man and nature, whether industrial or primitive. Although whaling is a rationalized, collective operation, based on a strict division of labor, it remains a bloody, murderous hunt. Playing this fact against the illusion that civilized man has won his freedom from physical nature, Melville transforms the line into an emblem of our animal fate. It signifies that we are bound to the whale by the needs and limitations of all living things, as, for example, hunger and death. Our precarious situation is illustrated by a man sitting in a whale boat. Technological progress does not alter the fact that his life remains enveloped in whale lines; at any moment the line may snatch him out of the boat. "For, when the line is darting out, to be seated then in the boat, is like being seated in the midst of the manifold whizzings of a steam-engine in full play, when every flying beam, and shaft, and wheel, is grazing you."

Throughout *Moby-Dick* Melville uses machine imagery to relate the undisguised killing and butchery of whaling to the concealed violence of "civilized" Western society. In fact the primitive urge back of the machine is what seems to invest it with a sense of fatality. ("We have constructed a fate," Thoreau says of the railroad, "an *Atropos*, that never turns aside.") To Ishmael the line is at once a token of man's inescapable need to consume the whale and of the whale's deadly hold upon him. Even as it lies in graceful repose, the soft, golden-haired Manila spells fatal power: "as it silently serpentines about the oarsmen before being brought into actual play—this is a thing which carries more of true terror than any other aspect of this dangerous affair." The reason for terror becomes evident the first time Queequeg throws his harpoon. Here a "savage" is attacking a beast, yet as Ishmael describes the scene it presages the release of explosive, uncontrollable energy in the era of steam power:

> A short rushing sound leaped out of the boat; it was the darted iron of Queequeg. Then all in one welded commotion came an invisible push from astern, while forward the boat seemed striking on a ledge; the sail collapsed and exploded; a gush of scalding vapor shot up near by; something rolled and tumbled like an earthquake beneath us.

The whale-line, to be sure, is only one of many lines of fate that crisscross the world of *Moby-Dick*. The line reappears on the charts, in the sword-mat that strikes Ishmael as the Loom of Time, in the "monkey-rope" that ties him to Queequeg, on the brow ("pleated with riddles") of the whale and in the "crucifixion" on Ahab's ravaged face. In the end, however, the whale line is the decisive line. On the third day of the chase, after Ahab darts the harpoon toward the white whale, the line runs foul; he stoops to clear it, clears it, but as he does the flying turn catches him round the neck and silently whisks him out of the boat. "Next instant . . . the rope's final end flew out of the stark-empty tub, knocked down an oarsman, and smiting the sea, disappeared in its depths." [30]

A whaling captain's life, as Ishmael sees it, is peculiarly conducive to illusions of Promethean power. Ahab's vocation endows him with a mastery of what Carlyle had called the "machinery" of the age — in both senses of the word. As captain he is, of course, master of an intricate piece of machinery in the literal sense. More important, however, is the mechanistic habit of mind, which Carlyle had attributed to modern man and which enables Ahab to control the consciousness of his men. He assumes that their minds function according to simple, mechanical principles, which is to say that they are essentially passive before experience. In this respect they might be inhabitants of Thoreau's "Concord." When Ahab sums up the results of his quarter-deck speech ("my one cogged circle fits into all their various wheels, and they revolve") the metaphor defines the subtle, complementary relation between his mind and theirs: he supplies the motive power — the ends — for which they are the means. A similar premise underlies the ingenious "political" strategy he formulates in "Surmises" (Chapter 46).

Ahab's concern here is the crew's morale, or, as he sees it, keeping his human tools in order: "To accomplish his object Ahab must use tools; and of all tools used in the shadow of the moon, men are most apt to get out of order." His special problem is that he cannot hope to hold the men only to that "one final and romantic object" he had proclaimed on the quarter-deck. That speech had confirmed his power to inspire men, but Ahab is too experienced to imagine that he can sustain such fervor until he locates Moby Dick. "In times of strong emotion," he knows, "mankind disdain all base considerations; but such times are evanescent." How, then, to keep the men satisfied until the critical moment? Ahab's solution is based on the idea, advanced by Carlyle, Marx, Emerson, and Thoreau (among others), that the Age of Machinery transforms men into objects: "The permanent constitutional condition of the manufactured man, thought Ahab, is sordidness." Under the circumstances, he concludes, his best hope is to exploit the simple, quantitative, acquisitive system of value honored by a capitalist society. Hence the nailing of the doubloon to the mast; hence the decision to maintain a normal whaling routine until the last moment.

> I will not strip these men, thought Ahab, of all hopes of cash — aye, cash. They may scorn cash now; but let some months go by, and no prospective promise of it to them, and then this same quiescent cash all at once mutinying in them, this same cash would soon cashier Ahab.

Nowhere does Melville display greater political acumen than in defining the paradoxical affinity between Ahab and the mechanistic view of life. As the *Pequod* approaches the white whale, Ahab's preoccupation with power becomes obsessive. Images of machinery, iron, forges, wheels, fire, and smoke fill his speech. In a fit of manic inspiration he "orders" from the blacksmith, whom he calls Prometheus, a "complete man" fifty feet high with a chest modelled after the Thames Tunnel, no heart at all, and about a quarter of an acre of brains. But Ahab feels an affinity with the new power only as a means — an instrument of his own wild, metaphysical purpose. He does not share the cloudy vision of material happiness that is the ostensible goal of industrial progress in America and — though probably with fewer illusions — throughout Western society. He cares nothing for wealth or worldly satisfactions (he lacks "the low, enjoying power"), but he knows how to manipulate those who do care. His ability to complement their limited perceptions is a savage comment upon a culture obsessed with the technical, external arrangements of life. Ahab's "cogged circle" synchronizes with their various wheels because he provides what they so conspicuously lack: a passionate sense of common purpose. The shattering fact is that his madness fits the needs of this technically proficient society. "Now,

in his heart, Ahab had some glimpse of this, namely: all
my means are sane, my motive and my object mad." Ruthless insight combined with burning purpose is what enables Ahab to gain a total dominion over the officers, men,
and owners * of the *Pequod* — over everyone, finally, but
Ishmael.[31]

In addition to Ahab's monomania, which is in some
measure a product of the whaling life, Ishmael extends
his inquiry to the object of the vengeful quest — the larg-
est of living creatures, the great whale. Even as he surrenders to Ahab's oratory, he vows to explain the "mystical
and well-nigh ineffable" horror the whale arouses in him.
And he goes to the heart of the matter at once. "It was
the whiteness of the whale," he says, "that above all things
appalled me." In discussing that unusual color Ishmael
addresses the reader with offstage candor, not as an ordinary character, but in his rôle as a narrator, poet, and
symbol-maker whose only hope of making the quest credible is to account, first, for his own seemingly unaccount-
able response to whiteness: ". . . explain myself I must,"
he concedes, "else all these chapters might be naught."
Ishmael's analysis of the whiteness of the whale (Chapter
42) is a majestic rhetorical feat appropriate to a subject
of ultimate concern: his own effort to apprehend one
natural object as it exists in itself — to isolate the naked,
irreducible fact responsible for his fear. He takes the mysterious color of the whale as a token of all the objects of

---

* Toward the absentee owners he adopts an icily cynical attitude,
encouraging them to think that he shares their interest in cash. In
stating the case Melville anticipates the twentieth century's discovery that
prudential, rationalistic cultures are peculiarly vulnerable to leaders of
mad purpose: "They were bent on profitable cruises, the profit to be
counted down in dollars from the mint. He was intent on an audacious,
immitigable, and supernatural revenge."

knowledge pursued by men, and so his voyage of mind epitomizes the attempt to know the nature of nature. After cataloguing the associations of whiteness, positive and negative, he discovers that his response is attributable neither to beneficent nor to abhorrent associations. Indeed, he is unable to trace the terror he feels to any specific meaning, benign or repellent, or to any value, good or evil, and he finally is drawn to the conclusion that it is precisely the indefiniteness of this fact, its hospitality to contradictory meanings and values, that causes fear. And then he adds the reflection that all colors, even "the sweet tinges of sunset skies and woods," are not "actually inherent in substances, but only laid on from without; so that all deified Nature absolutely paints like the harlot, whose allurements cover nothing but the charnel-house within. . . ." In the end Ishmael decides that what strikes terror to the soul is not the object at all, but rather its unknowability. Is not the essence of whiteness, he asks, that "dumb blankness, full of meaning, in a wide landscape of snows — a colorless, all-color of atheism from which we shrink?" The ambiguity evoked by the white whale is what accounts for Ishmael's ineffable horror. And ambiguity is the attribute of the physical universe that matches the contradiction at the heart of a culture that would deify the Nature it is engaged in plundering.

All Ishmael's other efforts to apprehend the whale bear out these skeptical conclusions. In this respect his experience aboard the *Pequod* parallels Thoreau's experience at the pond. Ishmael is determined to know whales. He subjects the anatomy of the beast to a meticulous examination, but again and again he discovers, as Thoreau does, that the facts will not of themselves yield anything like a clear, unequivocal truth. After examining many pictures of whales he decides that "the great Leviathan is that one creature in the world which must remain unpainted to the last." No single portrait can possibly capture the truth, Ishmael says, and so "there is no earthly way of finding out precisely what the whale really looks like." But he persists. The chapters dealing with parts of the whale's anatomy begin, typically, with a straightforward attempt to set forth the evidence, but in the end he often admits that the mere facts do not suffice:

> The more I consider this mighty tail, the more do I deplore my inability to express it. At times there are gestures in it, which, though they would well grace the hand of man, remain wholly inexplicable. . . . Dissect him how I may, then, I but go skin deep; I know him not, and never will. But if I know not even the tail of this whale, how understand his head? much more, how comprehend his face, when face he has none?

When Ishmael's investigations of natural fact do not issue in confessions of failure they often turn, suddenly, into undisguised flights of poetic imagination. Like Thoreau, he discovers that only the symbol-making power of mind can induce a fact to flower into a truth.[32]

As the voyage continues and Ishmael becomes increasingly skeptical about the possibility of apprehending the whale, he is drawn into himself. For a time his voice is muted. But his skepticism and his growing inwardness express themselves, eventually, in the crisis of chapters 94 through 96. Here, for the second time, Melville uses a variant of the machine-in-the-garden design to compose an ironic juxtaposition of episodes. This time, however, he probes more deeply into the psychic significance of the contrast. Drawing insights from his own conflicting emotions, Melville expands the opposition between Ahab and Ishmael into an all-engulfing metaphor of contradiction.

In "A Squeeze of the Hand" (Chapter 94), Ishmael again abandons himself to the "all" feeling. With several other sailors he is seated before a tub filled with sperm oil; their job is to squeeze the hardening lumps of spermaceti back into fluid in preparation for the try-works. "A sweet and unctuous duty!" exclaims Ishmael, recalling the ancient reputation of "sperm" as a cosmetic — an incomparable clearer, sweetener, softener, mollifier. "After having my hands in it for only a few minutes," he says, "my fingers felt like eels, and began, as it were, to serpentine and spiralize." Sitting there all morning squeezing sperm induces a strange euphoria:

> As I sat there at my ease, cross-legged on the deck; after the bitter exertion at the windlass; under a blue tranquil sky; the ship under indolent sail, and gliding so serenely along; as I bathed my hands among those soft, gentle globules of infiltrated tissues, woven almost within the hour; as they richly broke to my fingers, and discharged all their opulence, like fully ripe grapes their wine; as I snuffed up that uncontaminated aroma, — literally and truly, like the smell of spring violets; I declare to you, that for the time I lived as in a musky meadow; I forgot all about our horrible oath; in that inexpressible sperm, I washed my hands and my heart of it; I almost began to credit the old Paracelsan superstition that sperm is of rare virtue in allaying the heat of anger: while bathing in that bath, I felt divinely free from all ill-will, or petulance, or malice, of any sort whatsoever.

Though in many ways like the masthead episode, this is a pastoral interlude with a difference. Here Melville leaves no doubt about the genesis of the "all" feeling. The fragrance of spring violets and the musky meadow figure in Ishmael's euphoria, it is true, but they exist wholly within his imagination. The external landscape is on the periphery, a source of remembered images, while the center of the experience is unabashedly erotic:

> Squeeze! squeeze! squeeze! all the morning long; I squeezed that sperm till I myself almost melted into it; I squeezed that sperm till a strange sort of insanity came over me; and I found myself unwittingly squeezing my co-laborers' hands in it, mistaking their hands for the gentle globules. Such an abounding, affectionate, friendly, loving feeling did this avocation beget; that at last I was continually squeezing their hands, and looking up into their eyes sentimentally. . . .

In describing this spell Melville makes his deepest penetration into the psychic sources of the "all" feeling — and of sentimental pastoralism generally. The profundity of his insight is indicated by the brilliance of the clarifications that follow. Here, for the first time, he actually explains the redemptive process hitherto evoked by Ishmael's "melting" figure. "I felt a melting in me," the narrator had said of the emotions initially aroused in him by his relation with Queequeg. "No more my splintered heart and maddened hand were turned against the wolfish world. This soothing savage had redeemed it." The basis of this feeling, it now appears, is what Freud no doubt would have called an infantile pleasure ego. Erotic fulfillment of a childish kind exposes the unshielded inner self in all its innocent yearning. This experience issues in a mild, submissive, feminine view of life, opposed in every way to Ahab's irremediable aggressiveness. And it entails a renunciation, or lowering, of objectives. For he has perceived, Ishmael says, "that in all cases man must eventually lower, or at least shift, his conceit of attainable felicity; not placing it anywhere in the intellect or the fancy; but in the wife, the heart, the bed, the table, the saddle, the fire-side, the country . . ." [33]

What is striking about this vision of rural tranquillity, under the circumstances, is its irrelevance. It is totally, absurdly, not to say insanely irrelevant to Ishmael's life, to the whaling enterprise or, in other words, to reality as defined by the governing metaphor of *Moby-Dick*. To forget the horrible oath is to forget the actual situation aboard the *Pequod*. On this ship only Ahab determines the effective "conceit of attainable felicity." This is not to say that Ishmael's moment of loving transport was unreal. It had been genuine enough. But what plays the mischief with the truth at such times, as Melville had explained in his letter to Hawthorne, is "that men will insist upon the universal application of a temporary feeling or opinion." Once again Melville places Ishmael on the verge of a hideous fall into the vortex of reality. The danger is hinted by the abrupt transition from the serene, sun-filled eroticism of "A Squeeze of the Hand" to the violent, firelit passions of "The Try-Works." The distance between these two episodes is the unbridgeable gulf between a childlike self, intoxicated by thoughts of innocent love, and the outer world of ruthless, adult aggressiveness. Between the two chapters Melville cunningly places "The Cassock" (Chapter 95), a brief Rabelaisian tribute to sexual energy as figured by the "grandissimus," or phallus, of the whale.

The try-works episode stands in the same relation to "A Squeeze of the Hand" as the quarter-deck scene to the masthead revery: it is an evocation of explosive power following a pastoral interlude. But here, presumably as a result of Ishmael's access to deeper levels of consciousness, Melville's narrator fuses into a single, compelling image the inward and outward implications of man's assault upon nature. The chapter begins, in a flat, documentary style, with an account of the technical apparatus. Ishmael describes the anomalous masonry "works," peculiar to American whalers, which look like nothing so much as a brick-kiln from an open field. Taken together, the details form a composite image of industrial technology in the Age of Steam. Ishmael sketches the elaborate iron-work, the iron mouths of the "furnaces," the heavy iron doors, and the water reservoir needed to preserve the deck from the intense heat of the fire; the scraps of dried blubber used for fuel (the whale, he notes, is "self-consuming") surround the whole operation in an unspeakable, stinking, horrible smoke. The factory aspect befits the rendering process, which is the industrial phase, after all, of the whaling business. It also enables Melville to fashion a metaphor of heedless, unbridled, nineteenth-century American capitalism. Extracting economic value from the whale is, in theory at least, the commanding, official motive of the journey. All else is secondary. The try-works provide an appropriate setting for the social action of *Moby-Dick* — the saga, as Thorstein Veblen might have put it, of a predatory culture.

As Ishmael continues, describing the scene on deck the first time the works were fired, the connotation shifts from civilized industry to a primitive rite. It is midnight. The pagan harpooners, who are always the "stokers," look like infernal shapes in the firelight. This fire illuminates the primal energies at work back of the modern, seemingly rational, drive for profit and power.

With huge pronged poles they pitched hissing masses of blubber into the scalding pots, or stirred up the fires beneath, till the snaky flames darted, curling, out of the doors to catch them by the feet. The smoke rolled away in sullen heaps. . . . Their tawny features, now all be-grimed with smoke and sweat, their matted beards, and the contrasting barbaric brilliancy of their teeth, all these were strangely revealed in the capacious emblazonings of the works.

The men now tell each other their "unholy adventures," and the excitement rises. What had begun as a routine whaling operation becomes a weird, violent, orgiastic rite. The scene provides a glimpse into both the past of West-ern civilization and its unconscious present. Almost im-mediately, Melville shifts the focus again, the wild fire-ship now evoking Ahab's inner life — the frenzied, walled-in passion of an American captain "damned in the midst of Paradise!"

As they narrated to each other their unholy adventures, their tales of terror told in words of mirth; as their uncivilized laughter forked upwards out of them, like the flames from the furnace; as to and fro, in their front, the harpooners wildly gesticulated with their huge pronged forks and dippers; as the wind howled on, and the sea leaped, and the ship groaned and dived, and yet steadfastly shot her red hell further and further into the blackness of the sea and the night, and scornfully champed the white bone in her mouth, and viciously spat round her on all sides; then the rushing Pequod, freighted with savages, and laden with fire, and burning a corpse, and plunging into that blackness of darkness, seemed the material counterpart of her monomaniac commander's soul.

That night a strange thing happens to Ishmael. Standing at the helm, watching the "fiend shapes" capering in the smoke and fire, he is suddenly aware of something fatally wrong. A stark, bewildered feeling, "as of death," comes over him. Reaching convulsively for the tiller, he realizes in the nick of time that he had unconsciously turned himself around, and that in another moment he might have capsized the ship. The incident is reminiscent of "Ethan Brand," and of the Faustian quest which had been inspired, initially, by a fiend evoked from the fires of a brick kiln. In *Moby-Dick*, of course, the monomaniac committed to a fiery quest is Ahab. From the beginning, however, Ishmael has acknowledged similar impulses in himself. And in this episode he comes closest to total capitulation. "Wrapped, for that interval, in darkness myself, I but the better saw the redness, the madness, the ghastliness of others." While under the spell Ishmael very nearly causes the catastrophe to which Ahab's monomania will lead.

But he comes back to his senses in time, and the upshot is an explicit, rational repudiation of the quest — not to be confused with the "forgetting" that had accompanied the "all" feeling. Now, in glad, grateful relief from the "unnatural hallucination of the night," he steps forth and addresses us directly; he is a teacher pointing the lesson in the manner of an Old Testament prophet: "Look not too long in the face of the fire, O man! Never dream with thy hand on the helm!" Then he enlarges upon the fire-sun antinomy:

> . . . believe not the artificial fire, when its redness makes all things look ghastly. To-morrow, in the natural sun, the skies will be bright; those who glared like devils in the forking flames, the morn will show in far other, at

least gentler, relief; the glorious, golden, glad sun, the
only true lamp — all others but liars!

In paying tribute to the sun, however, Ishmael does not
regress to the sunlit dream world of "A Squeeze of the
Hand." Nor does he reaffirm the "conceit" which had
caught his fancy in that intoxicated moment when, having
washed away the oath in inexpressible sperm, he had
"lived as in a musky meadow." The vision he now gains is
illuminated by the "natural sun," a mode of seeing that is
not, like the other, a half-sight. It does not mean averting
one's eyes from the dark side, from the suffering to which
Ahab's grand, angry quest testifies.

> Nevertheless, the sun hides not Virginia's Dismal
> Swamp, nor Rome's accursed Campagna, nor wide Sa-
> hara, nor all the millions of miles of deserts and of griefs
> beneath the moon. The sun hides not the ocean, which
> is the dark side of this earth, and which is two thirds
> of this earth.

At this point Melville moves his narrator toward a saving resolution. If two-thirds of this earth is a dark, woeful,
hideous wilderness, then, Ishmael reasons, the man who
has more of joy than sorrow in him cannot be a true
man. (A true man is one whose responses match the
actual condition of things.) With this perception Ishmael
finally has outgrown his proclivity to childish pleasure
fantasies. The true man, he concludes, is the man of
sorrows, and the truest book, Ecclesiastes, cuts with an
edge that has been hammered fine by woe. Ahab, accordingly, is a truer man than the Ishmael who dwelled
upon "watery pastures" and "musky meadows." And there
is a large truth, a two-thirds truth, as it were, in the aggressive Ahab whose sign is the doomed American whaler,

freighted with savages, laden with fire, burning a corpse, and plunging into blackness. But it is only a partial truth, and here Ishmael takes the final turn toward a third, rational, and more complex position. The conclusion of "The Try-Works" is at once a repudiation of the quest, a reaffirmation of reason, and a tribute to the man large enough to withstand the extremes of hope and fear:

> But even Solomon, he says, "the man that wandereth out of the way of understanding shall remain" (*i.e.* even while living) "in the congregation of the dead." Give not thyself up, then, to fire, lest it invert thee, deaden thee; as for the time it did me. There is a wisdom that is woe; but there is a woe that is madness. And there is a Catskill eagle in some souls that can alike dive down into the blackest gorges, and soar out of them again and become invisible in the sunny spaces. And even if he for ever flies within the gorge, that gorge is in the mountains; so that even in his lowest swoop the mountain eagle is still higher than other birds upon the plain, even though they soar.

After Ishmael has described virtually every feature of the whale, he comes at last to the question of its innermost structure; and here, in "A Bower in the Arsacides" (Chapter 102) and "Measurement of the Whale's Skeleton" (Chapter 103), Melville uses a third variant of the pastoral design. It behooves me, Ishmael says, to "set him before you in his ultimatum . . . his unconditional skeleton." Whereupon he tells of an adventure he had had many years before the voyage of the *Pequod*.

Having been invited by Tranquo, king of Tranque (a land of Melville's invention), to spend a holiday in a seaside glen, he sees among the treasures of this primitive people a skeleton of a great sperm whale. It has been moved from the beach where it was found to a lush green glen, where the priests had converted it into a temple.

> It was a wondrous sight. The wood was green as mosses of the Icy Glen; the trees stood high and haughty, feeling their living sap; the industrious earth beneath was as a weaver's loom, with a gorgeous carpet on it, whereof the ground-vine tendrils formed the warp and woof, and the living flowers the figures.

Here inside the whale Ishmael is as close to the center of things as he is likely to get. The skeleton is a primitive temple of nature. The priests keep a holy flame burning within, and the "artificial smoke" pours out of the hole where the real watery jet once had come forth.

As Ishmael moves further into the whale, however, the image shifts. The bones of the skeleton are crisscrossed with vines, and through them the sunlight seems "a flying shuttle weaving the unwearied verdure." All at once the intricate structure of bones, the smoke pouring out, the light working through the lacing of leaves — all combine to give him the impression that he is inside a textile factory. The image makes possible a metaphoric flight. Wherefore, asks Ishmael, "these ceaseless toilings? Speak, weaver! — stay thy hand! — but one single word with thee!" Pursuing the analogy between human and natural productivity, he in effect asks: what is the ceaseless striving for? why the endless production and reproduction? But the weaver ignores his demand:

> Nay — the shuttle flies — the figures float from forth the loom; the freshet-rushing carpet for ever slides away. The weaver-god, he weaves; and by that weaving is he deafened, that he hears no mortal voice; and by that hum-

ming, we, too, who look on the loom are deafened; and only when we escape it shall we hear the thousand voices that speak through it. For even so it is in all material factories.

Here, growing in the whale's skeleton, is the greenness Ishmael has been seeking, yet that same greenness has the aspect of a factory. It is a bold conceit, this green factory inside the whale — another vivid metaphor of American experience: Ishmael deliberately making his way to the center of primal nature only to find, when he arrives, a premonitory sign of industrial power. Art and nature are inextricably tangled at the center. Hence there is no way to apprehend the absolute meaning of a natural fact, no way to seize the whale in his "ultimatum," that is, as he exists in himself, independent of a human perceiver. How vain and foolish, says Ishmael, to try "to comprehend aright this wondrous whale" by poring over his dead skeleton "stretched in this peaceful wood." The answer is not to be found by measuring specimens. "No. Only in the heart of quickest perils; only when within the eddyings of his angry flukes; only on the profound unbounded sea, can the fully invested whale be truly and livingly found out."

In the heart of the perils aboard the *Pequod* Ishmael rediscovers the green fields. But they do not signify what he had allowed himself to imagine on the masthead or while squeezing spermaceti. On those occasions his sense of reality had been restored by Ahab and his fiery hunt; after repeated experiences he had learned that the existence of greenness does not justify the sentimental "all" feeling. On the other hand, it does provide a warrant for rejecting Ahab's quest. In fact, the captain's mad purpose results from an inversion of the same impulse that gives rise to that sentimental state. Ahab also would make a universal application of a limited experience. Toward the climax of the voyage, accordingly, Ishmael adopts a more consistently skeptical attitude toward external nature. He continues to enjoy moments of "dreamy quietude" (as in "The Gilder," Chapter 114), when, he admits, a "filial, confident, land-like feeling towards the sea" has come over him, and the ocean reminds him of "grassy glades," and "ever vernal . . . landscapes in the soul"; but he reminds himself of the danger inherent in such feelings, and that "when beholding the tranquil beauty and brilliancy of the ocean's skin, one forgets the tiger heart that pants beneath it." Here Ishmael is defining a complex pastoralism, a view of experience that matches the duality of nature. The possibility of joyous fulfillment exists; the image of green fields is meaningful, but only so far as it is joined to its opposite. In itself the image represents neither a universal condition nor a set of values which can be embodied in social institutions. In many ways Ishmael's point of view is akin to Thoreau's at the end of *Walden*. In his rôle as ordinary seaman, therefore, Ishmael virtually drops out of sight. The figure who is saved from the wreck is the narrator and symbol-maker of *Moby-Dick;* he is Job's messenger.

But Melville does not exaggerate the saving power of this complex, literary vision. It may account for the rescue of one man, but it is powerless to save the ship. Only Starbuck tries to divert Ahab, and his failure dramatizes the weakness of sentimental pastoralism. Starbuck is a prudent man and temporizer whose very name suggests idealism joined with the "sordidness" of the cash-minded manufactured man. Although he fancies himself a realist, Starbuck

cannot face the brutal facts. In the same episode in which
Ishmael recalls the tiger heart panting beneath the vernal
landscapes of the soul, Starbuck, "gazing far down from
his boat's side into that same golden sea," murmured:

> "Loveliness unfathomable, as ever lover saw in his
> young bride's eye! — Tell me not of thy teeth-tiered
> sharks, and thy kidnapping cannibal ways. Let faith oust
> fact; let fancy oust memory; I look deep down and do
> believe."

Until the end Starbuck clings to the notion that Ahab
can be deterred by a comparably sentimental faith. But
that fallacy is exposed in "The Symphony," the last chapter before the sighting of the white whale, where Melville
finally clarifies the relation between greenness and Ahab's
monomania.[34]

What misleads Starbuck in this episode, quite apart
from his habitual tendency to deny the cannibal under-
side of reality, is Ahab's momentary response to a lovely,
clear, calm, steel-blue day on the Pacific. It is one of those
days when the air seems "transparently pure and soft,
with a woman's look," and when it is easy to forget the
mighty leviathans, sword-fish, sharks — the "strong, troubled, murderous thinkings of the masculine sea" — rushing to and fro down in the bottomless ocean. Noticing that
Ahab is deeply affected, as if the mildness and beauty of
the day had dispelled the "cankerous thing" in his soul,
Starbuck approaches him. The captain turns and utters
what can only be described as Ahab's elegy for Ahab. It is
his final effort to explain himself. The dominant theme
of the lament is separation from greenness. First, Ahab
summarizes his forty years at sea, forty years of privation,
peril, and storm-time. To a remarkable degree his obses-

sions are traceable to his vocation. As he describes it, his life stands forth as a pathological instance of the traits nurtured by a society obsessed with accumulation, competitive performance, and power. Ahab's mind has been formed by the chase, a relation to "otherness" (or nature) that is almost wholly aggressive. The whaling life has maimed him, "unmanned" him, and provoked his mad commitment to an unattainable, transcendent purpose. Earlier, on the quarter-deck, he had described the whale as a wall shoved near; and now, speaking to Starbuck, he pictures himself as a lifelong prisoner within the " 'walled-town of a Captain's exclusiveness, which admits but small entrance to any sympathy from the green country without . . .' " Here, drawing together the pastoral imagery developed by Ishmael, Melville imparts an astonishing range of implication to Ahab's elegy. Estrangement from greenness is symptomatic at once of a collective and an individual illness. When Ahab considers the "dry salted fare" he has eaten, "fit emblem of the dry nourishment" of his soul, and the facts about his marriage (he had married past fifty and sailed the next day), and the mad frenzy of the endless chase, he concedes that he has become " 'more a demon than a man! . . . a forty years' fool — fool — old fool. . . .' "

In "The Symphony" Melville joins the psychological, political, and metaphysical explanations of Ahab's quest. A whaling society, dedicated to an unbridled assault upon physical nature, selects and rewards men who adapt to the demand for extreme repression. But here, in "The Symphony," Ahab's repressed feelings suddenly rise to the surface. A wave of tenderness overcomes him. " 'By the green land,' " he says, gazing fondly into Starbuck's eyes, " 'by the bright hearth-stone! this is the magic glass, man; I see my

wife and my child in thine eye.' " Whereupon he orders
the mate to stay aboard when he lowers to kill Moby Dick.
Now Starbuck becomes elated; he urges the captain not
to pursue the "hated fish," but to fly from these "deadly
waters" back to Nantucket, where, he says, they also have
such mild blue days. But Starbuck again has allowed
hope to oust fact; he has mistaken a temporary emotion
of Ahab's for a permanent change of heart. (He is a perfect embodiment of the popular American notion that the
drive for power can be assuaged by appeals to sentiment.)
A moment later Melville establishes, once and for all, the
impotence of Starbuck's faith. Ahab cannot turn back. He
acknowledges that the quest stems from motives he can
neither understand nor control.

> "What is it, what nameless, inscrutable, unearthly
> thing is it; what cozzening, hidden lord and master, and
> cruel, remorseless emperor commands me; that against
> all natural lovings and longings, I so keep pushing, and
> crowding, and jamming myself on all the time; recklessly
> making me ready to do what in my own proper, natural
> heart, I durst not so much as dare? Is Ahab, Ahab? Is it
> I, God, or who, that lifts this arm? . . ."

Ahab's compulsion arises in troubled depths of mind
which are the psychic equivalent of the shark-like cannibalism of the sea. (At one point, here, he jocularly refers to
himself as "cannibal old me.") He is bound to the whale
by an invisible line, ghostly counterpart to the Manila
rope that binds mankind in a predatory relation to nature. By the time of "The Symphony" it is too late to
break through the wall that separates Ahab from greenness. Lacking "the low, enjoying power," subtly, malig-
nantly damned in the midst of paradise, his thoughts of
love turn to thoughts of death. "The Symphony" ends with

Ahab's meditation on death, "rust amid greeness," *Et in Arcadia Ego*. Before he is finished, Starbuck, "blanched to a corpse's hue with despair," steals away. The next morning, at dawn, Ahab sights the white whale and the three-day chase begins. It is fitting that the prelude to Melville's American apocalypse should be an elegiac pastoral interlude directed against a sentimental trust in nature.

The closing action of *Moby-Dick* brings an impersonal, dramatic authority to Ishmael's idea of the whale, and of man's relation to it. Moby Dick is neither good nor evil. When he slowly rises out of the water, his marbleized body for an instant forming "a high arch, like Virginia's Natural Bridge," he seems a grand god revealing himself. His attributes invite contradictory interpretations like those evoked by the unspoiled terrain of the New World. At first a "gentle joyousness — a mighty mildness of repose in swiftness" invests the gliding whale. But later, after three days of pursuit, he seems quite another creature: "Retribution, swift vengeance, eternal malice were in his whole aspect. . . ." The encounter between the *Pequod* and the whale is cataclysmic, yet Melville's imagery tends to break down the usual distinction between art and nature. There is an almost mechanical predictability about Moby Dick's behavior, and Ahab is able to foretell his course from day to day as if it had been rationally plotted. "And as the mighty iron Leviathan of the modern railway is so familiarly known in its every pace, that, with watches in their hands, men time his rate . . . even so, almost, there are occasions when these Nantucketers time that other Leviathan of the deep. . . ." But if there is a principle of order akin to scientific rationality in the whale, there is a wild, sharkish turbulence astir in Ahab's mind. "See!" says Starbuck on the third day, "Moby Dick seeks thee not. It is thou, thou, that madly seekest him!" But no perception of the natural facts, no plea for survival, no appeal to the heart, can touch Ahab. He is the perverted, monomaniac incarnation of the Age of Machinery, as he himself had admitted earlier: "all my means are sane, my motive and my object mad."[35]

When the last chip of the *Pequod* is carried, spinning round and round, into the vortex, Ishmael is in the water. Shortly before he had been flung from Ahab's boat, "helplessly dropping astern, but still afloat and swimming." And when the white buttress of the whale's forehead had smashed into the ship, Ishmael was floating on the margin of the scene, yet within full sight of it. In fact he had been "drawn towards the closing vortex," almost suffering the fate he had so narrowly escaped in the mast-head and try-works episodes. Just as Ishmael reached the "vital centre" of the foaming pool, the black bubble burst and released the coffin life-buoy that kept him afloat on the soft and dirge-like main. So Ishmael is saved. He has relocated greenness, shifting it from the green fields of the Republic to a necessary, but by no means sufficient, principle of survival. In accomplishing Ishmael's "salvation," Melville in effect puts his blessing upon the Ishmaelian view of life: a complex pastoralism in which the ideal is inseparably yoked to its opposite. It is a doctrine that arises at the "vital centre" of experience. At the same time Melville acknowledges the political ineffectuality of this symbol-maker's truth. Ishmael survives, to be sure, but as an "orphan" floating helplessly on the margin of the scene as society founders. Of the qualities necessary for survival, Melville endows Ahab with the power and

Ishmael with the wisdom. Ishmael is saved as Job's messengers had been saved, in order that he may deliver to us a
warning of disasters to come.

4

No book confirms the relevance of the pastoral design to
American experience as vividly as the *Adventures of
Huckleberry Finn* (1884). Here, for the first time, the
mode is wholly assimilated to a native idiom. Most of the
unique features of the novel follow from the inspired idea
of having the hero tell his own story in his own language.
By committing himself to a vernacular narrator, Samuel
Clemens lends the sense of place a freshness and lyricism
unmatched in American writing. In this case, however, the
presence of the machine-in-the-garden motif has as much
to do with the defects as with the merits of the work. In
Huckleberry Finn the sudden incursion of the new power
has the effect of a shattering childhood trauma; it is an
aspect of a pattern of contradiction so nearly absolute
that the promised resolution of comedy is unconvincing,
and as a result Clemens's ingenious efforts to contrive an
affirmation virtually break his masterpiece in two.

That *Huckleberry Finn* initially was inspired by an affirmative impulse cannot be doubted. On the first page
Clemens explains that the book is a sequel to his comic
idyll of boyhood, *The Adventures of Tom Sawyer* (1876).
In fact, the surge of literary activity of which *Huckleberry
Finn* is the peak had begun, in 1874, as a deliberate effort
to reclaim the past. What had set Clemens off at the time
was the idea of recovering (as he puts it in a letter to William Dean Howells) the "old Mississippi days of steam-
boating glory & grandeur as I saw them (during 5 years)

from the pilot house." With that object in view, he wrote the series of reminiscences, "Old Times on the Mississippi," which ran in the *Atlantic Monthly* in 1875 and later was incorporated in *Life on the Mississippi* (1883).[36]

Once the well of nostalgia had been opened, it kept filling. During the next ten years Clemens experienced a release of imaginative energy comparable to Melville's "unfolding" between *Typee* and *Moby-Dick*. The success of "Old Times" encouraged him to write *Tom Sawyer*, and then, still riding the buoyant flood of memory, he began *Huckleberry Finn*. At first the work went smoothly. During July and August of 1876 he completed roughly four hundred manuscript pages (about a fourth of the novel), up to the point at the end of Chapter 16 where the steamboat smashes the raft. But there he reached an impasse. He put the project aside for about three years; then, beginning in 1879, he worked on it intermittently, but did not complete it until 1883. His difficulty may be described in several ways: it was at once a problem of attitude and of technique, meaning and structure, and, as he himself had defined it in "Old Times," it arose from a conflict at the center of American experience.[37]

The theme of "Old Times" is "learning the river." Here the narrator, Mark Twain, tells of his initiation into a unique American vocation: Mississippi piloting. He insists that this job has to be learned by an apprentice on the spot; no books, no school, no theory can equip him. What he has to learn is a new language — nothing less than the language of nature. It is not simply the general technique of piloting, but a particular piece of geography which he must possess. He must "know the river" by day and by night, summer and winter, heading upstream and heading downstream. He must memorize the landscape. And this knowledge forever distinguishes him from the uninitiated. To the trained pilot the river is a book that delivers its most guarded secrets; but to ignorant passengers it is only a pretty picture. Learning the river is an exhilarating experience, and in re-creating it Clemens manages to impart that sense of freedom and power to his writing. (By comparison, the second volume of *Life on the Mississippi*, which lacks this theme, is dull and labored.) Yet — and here a problem arises — Mark Twain admits that in acquiring the new lore he loses something too: nothing less, in fact, than his sense of "the grace, the beauty, the poetry" of the majestic river. It is gone. In learning the matters of fact necessary to his vocation the pilot loses, or so he thinks, the capacity to enjoy the beauty of the landscape.[38]

By way of illustration, he compares two ways of experiencing a sunset on the river. First, he describes the brilliant sunset as he once might have enjoyed it, in his innocence:

> A broad expanse of the river was turned to blood; in the middle distance the red hue brightened into gold, through which a solitary log came floating, black and conspicuous; in one place a long, slanting mark lay sparkling upon the water; in another the surface was broken by boiling, tumbling rings, that were as many-tinted as an opal; where the ruddy flush was faintest, was a smooth spot that was covered with graceful circles and radiating lines, ever so delicately traced; the shore on our left was densely wooded, and the somber shadow that fell from this forest was broken in one place by a long, ruffled trail that shone like silver; and high above the forest wall a clean-stemmed dead tree waved a single leafy bough that glowed like a flame in the unobstructed splendor that was flowing from the sun. There were graceful curves, reflected images, woody heights, soft dis-

tances; and over the whole scene, far and near, the dissolving lights drifted steadily, enriching it every passing
moment with new marvels of coloring.

What he describes is in fact a landscape painting in the
received picturesque tradition: soft distances, dissolving
lights, and graceful curves. In much of Clemens's early
work we find landscapes similarly composed, noble pictures seen through a "Claude glass." For instance, Venice,
in *Innocents Abroad* (1869), is like "a beautiful picture
— very soft and dreamy and beautiful." Or of Lake Tahoe,
in *Roughing It* (1872), we are told that a "circling border
of mountain domes, clothed with forests, scarred with
landslides, cloven by canyons and valleys, and helmeted
with glittering snow, fitly framed and finished the noble
picture." In these passages Clemens is working within an
established literary convention; he recognizes its shortcomings, and he often gives it a comic turn. He carries a
description of a Mediterranean vista in *Innocents Abroad*
to grand rhetorical heights, then abruptly cuts it off with:
"[Copyright secured according to law.]" Or a similar trick,
in *Roughing It,* follows an account of a "majestic panorama," when the narrator stops to eat, and "nothing," he
says, "helps scenery like ham and eggs." As a writer of
comic travel books, Clemens recognized that the stock
language of landscape appreciation was easily deflated by
exposure to simple, everyday facts. But if that style was
ludicrous, how was a writer to convey a genuine feeling
for scenery? What was the alternative to this elevated
rhetoric? [39]

The sunset in "Old Times" is a dramatization of this
literary problem. Before "learning the river," he says, the
pilot had enjoyed the stock response. Seeing boughs that
"glowed like a flame" and trails on the water that "shone

like silver," he had stood "like one bewitched," drinking it in, "in a speechless rapture." The calculated triteness of the language, a literary equivalent to the painter's picturesque, matches a trite view of nature. But that was the pilot's feeling before his initiation. Afterwards,

> . . . if that sunset scene had been repeated, I should have looked upon it without rapture, and should have commented upon it, inwardly, after this fashion: "This sun means that we are going to have wind to-morrow; that floating log means that the river is rising, small thanks to it; that slanting mark on the water refers to a bluff reef which is going to kill somebody's steamboat one of these nights . . . ; those tumbling 'boils' show a dissolving bar and a changing channel there; the lines and circles in the slick water over yonder are a warning that that troublesome place is shoaling up dangerously; that silver streak in the shadow of the forest is the 'break' from a new snag, and he has located himself in the very best place he could have found to fish for steamboats . . ."

And so on. The beautiful view, it seems, is for those who see only the surface of nature. Behind every appearance of the beautiful there is a fact of another sort. "No," says the initiated pilot, "the romance and the beauty were all gone from the river."

In spite of the false note here, a theatricality he found difficult to resist, Clemens has located a real problem. And it is more than a problem of language. As he was to demonstrate, he longs to convey a joyous, even reverent feeling for the landscape of the Mississippi Valley as he had known it in his youth. Yet neither of the available modes — the passengers' or the pilot's — will serve his purpose. What is more, the two ways of apprehending the river represent

opposed sets of values. One is identified with the passengers
and the uninitiated, novice pilot, the other with the older,
wiser narrator, Mark Twain. There are many differences
between these two points of view, but the most important
is the relation to nature each implies. The passengers are
strangers to the river. They lack the exact, technical
knowledge possessed by the trained pilot. Yet as spectators,
schooled in romantic landscape painting, they know what
to look for. They enjoy the play of light on the water.
Their aesthetic response, given the American geography,
Clemens associates with the effete, cultivated, urban, privileged East. The pilot, on the other hand, is of the West,
and in his calling he can scarcely afford to admire the
break in the shadow that shines like silver. To him it is
not merely a felicitous detail in a misted, beautiful picture. Responsible for the boat's safety, he must probe for
the real meaning of the silver streak: it is a "break" from
a new snag that might destroy a steamboat. (The snag
should be noted; it will reappear.) To do his job the pilot
must keep his mind upon the menacing "reality" masked
by the beautiful river.

The pilot's dilemma is a variant of a theme central to
Walden and *Moby-Dick,* and indeed it is omnipresent
in nineteenth-century culture. Like Melville's Ishmael,
Clemens's narrator is confronted with an impossible choice
between two modes of perception: one is aesthetically and
emotionally satisfying, yet illusive; the other is analytically
and practically effective, copes with harsh realities, yet is
devoid of all but utilitarian value and meaning. The
choice is between the mawkish sentiments of the passengers
and the bleak matter-of-factness of the pilot. For Samuel
Clemens, to be sure, this was not an abstract issue. Yet
neither was it simply another theme that he could ma-

nipulate as a cool, professional craftsman. For him the problem had an immediate, practical urgency: it was a problem of language, of style. If *Huckleberry Finn* is any indication, what Clemens wanted was to affirm the values embodied in the landscape *in its actuality*. He wanted to convey both the passengers' delight and the pilot's perception. He had encountered the problem as a writer, and it was as a writer (he was no theorist) that he finally came to grips with it. Beginning with "Old Times," he attempted in various ways to infuse the old mode with fresh feeling. "I perhaps made a mistake," he remarked to Howells after finishing *Tom Sawyer*, "in not writing it in the first person." Clemens's solution, if that is the correct word, is implicit in the choice of Huckleberry Finn as narrator of the sequel.[40]

Although published in 1884, *Huckleberry Finn* is set in the period of Hawthorne's Sleepy Hollow notes, on the eve of the take-off into rapid industrialization. On the title-page of the first edition, Clemens announces that he is taking a backward look:

SCENE: THE MISSISSIPPI VALLEY.
TIME: FORTY TO FIFTY YEARS AGO.

Except for this distinctively retrospective cast (Clemens was nine years old in 1844), the organizing design is the familiar one. Like *Walden* or *Moby-Dick*, *Huckleberry Finn* begins with the hero's urge to withdraw from a repressive civilization. After an intolerably long, quiet, dull evening with the Widow Douglas and Miss Watson, Huck retires disconsolately to his room and sits by the window. Here, in a hauntingly ambiguous passage, Clemens foreshadows the boy's movement from society toward nature. He is in a house filled with people who presumably care for him, yet he is depressed. His attention is directed toward the dark woods, as if something out there were trying to communicate with him.

> I felt so lonesome I most wished I was dead. The stars was shining, and the leaves rustled in the woods ever so mournful; and I heard an owl, away off, who-whooing about somebody that was dead, and a whippowill and a dog crying about somebody that was going to die; and the wind was trying to whisper something to me and I couldn't make out what it was, and so it made the cold shivers run over me. Then away out in the woods I heard that kind of a sound that a ghost makes when it wants to tell about something that's on its mind and can't make itself understood. . . . [41]

Later Huck is impelled toward the source of this wordless message. The brutality of his father, after the genteel tyranny of Miss Watson, compels him to forsake the benefits of civilization. He flees to Jackson's Island, where he is joined by Jim, and everything is "dead quiet," and life is lazy and comfortable. For a time the island is another of those enchanting pastoral oases so endlessly fascinating to the American imagination.* After a rise in the river, Huck and Jim are able to penetrate the innermost recesses of the island. Clemens describes their dreamlike journey, canoeing among the trees, in a passage reminiscent of the

---

* It is a mistake to think of Huck as an expositor of primitivist values. He has no illusions about noble savagery. To get away from constraint, boredom, and cruelty he moves in the direction of, but by no means all the way to, artless nature. He is appalled when Jim tells him that he has been living on berries. Untouched by romantic pieties about Nature, Huck carries along as much equipment as possible when he escapes from Pap's cabin: a frying-pan, a coffee-pot, tin cups, a knife, fishhooks, and a gun. Notice, too, that the raft is made of cut lumber, not logs.

vernacular tradition in American landscape painting, and especially of Edward Hicks, who in the 1840's painted many versions of his popular work "The Peaceable Kingdom":

> Daytimes we paddled all over the island in the canoe. It was mighty cool and shady in the deep woods even if the sun was blazing outside. We went winding in and out amongst the trees; and sometimes the vines hung so thick we had to back away and go some other way. Well, on every old broken-down tree, you could see the rabbits, and snakes, and such things; and when the island had been overflowed a day or two, they got so tame, on account of being hungry, that you could paddle right up and put your hand on them if you wanted to; but not the snakes and turtles — they would slide off in the water.[42]

The idyll is interrupted when Huck discovers that the villagers are about to search the island, and as a result he and Jim begin their downstream journey by raft. But the idyll is not ended: the raft becomes a mobile extension of the island. The fugitives continue to enjoy many of the delights they had known earlier, above all a sense of the bounty, beauty, and harmony made possible by an accommodation to nature. Floating quietly downstream, Huck and Jim attune themselves to the serene rhythm of the great river.

> We catched fish, and talked, and we took a swim now and then to keep off sleepiness. It was kind of solemn, drifting down the big still river, laying on our backs looking up at the stars, and we didn't ever feel like talking loud. . . . We had mighty good weather, as a general thing, and nothing ever happened to us at all, that night, nor the next, nor the next.

The image of the quietly drifting raft is a realization of freedom born of sensuous delight and a liberation of instinct. "Other places do seem so cramped up and smothery," Huck says, "but a raft don't. You feel mighty free and easy and comfortable on a raft." This rudimentary society of two, one black and one white, is an American Arcadia, an egalitarian wish-image: "what you want, above all things, on a raft, is for everybody to be satisfied, and feel right and kind towards the others." The river partly insulates Huck and Jim from the hostile world of the shore. Yet Clemens is realistic enough, and faithful enough to the logic of his ruling metaphor, to admit the limitations of the raft. It lacks power and maneuverability. It can only move easily with the current, that is, southward into slave territory.[43]

The powerlessness of the raft comports with the unshielded innocence of its passengers. As the fugitives drift southward Clemens builds up the contrast between their affectionate relations and the brutality of a society based on slavery. When the raft approaches Cairo, Illinois, the confluence of the Mississippi and Ohio Rivers where Huck and Jim had planned to head north toward freedom, the tension rises. At the same time Clemens grants the boy greater understanding. At the end of Chapter 15, after having played a cheap trick on Jim (a trick predicated upon the slave's mental inferiority), Huck makes an affecting apology. And when Jim describes his plan to "steal" his family, Huck suddenly realizes the subversive implications of his situation. "It most froze me to hear such talk," he says. One part of his mind, his "conscience," is indoctrinated with the values of the dominant white culture — but only one part, and shortly afterwards, when faced with the immediate prospect of Jim's capture, he scarcely hesitates before lying to save his friend. Even as he reaffirms the code, the inadequacy of the raft as a refuge from the aggressive adult world becomes apparent. Huck discovers that during the night, in the fog, they had floated past Cairo. (This fog, incidentally, is another instance of the ambiguous influence of the landscape: in this case it literally obscures the road to freedom.) With the river carrying the raft deeper into slave territory, the eventual failure of the quest seems inevitable. The sense of impending tragedy is quickened, the following night, by the ominous sound of a steamboat bearing down on them. "We could hear her pounding along," Huck explains, "but we didn't see her good till she was close." And she didn't sheer off:

> . . . all of a sudden she bulged out, big and scary, with a long row of wide-open furnace doors shining like red-hot teeth, and her monstrous bows and guards hanging right over us. There was a yell at us, and a jingling of bells to stop the engines, a pow-wow of cussing, and whistling of steam — and as Jim went overboard on one side and I on the other, she come smashing straight through the raft.

Here, at the end of Chapter 16, Clemens reached the impasse that delayed work on the book for three years. He faced a number of difficulties. Having set out to write a funny, episodic sequel to *Tom Sawyer,* he had allowed himself to develop the theme of Jim's escape from slavery. Simply from a logical point of view, that idea was not easy to reconcile with his initial plan: a picaresque journey southward to the lower Mississippi. Then, too, as Henry Nash Smith has demonstrated, he had begun to discover unsuspected depths of compassion and loyalty in the boy — emotional resources that exceeded the requirements of the comic convention. But all of these difficulties were in

fact aspects of one over-riding problem. Clemens had begun to realize that the voyage, so far as it involved a search
for a new and freer life, could not succeed. As he describes
it, freedom aboard the raft signifies much more than the
absence of slavery in the narrow, institutional sense. It
embraces all of the extravagant possibilities of sufficiency,
spontaneity, and joy that had been projected upon the
American landscape since the age of discovery. The
thought that this great promise was to be submerged in
history (in his view a dreary record of man's lost hopes)
gave rise to the image of a monstrous steamboat that suddenly bulged out of the night, big, scary, inexorable, and
smashed straight through the raft.* [44]

When Clemens resumed work on the book, three years
later, he still had found no satisfactory way out of the
quandary. His chosen mode, which called for a light-hearted
resolution, hardly was an appropriate vehicle for all that
was implied by the smashing of the raft. But he did contrive various ways of by-passing the difficulty. Most important, he shifted attention from Jim's flight, as the
unifying theme, to a satiric attack upon the hypocrisy of
Southern life. Again and again, in this allegedly Christian,
aristocratic, cultivated society, Huck meets unspeakable

---

   * Clemens makes even more explicit his sense of the opposition between
the raft and the forces represented by machine technology in *A Tramp
Abroad* (1880), one of the books he wrote during the hiatus in the composition of *Huckleberry Finn:* "The motion of a raft," he says, describing a journey down the Neckar, ". . . is gentle, and gliding, and smooth,
and noiseless; it calms down all feverish activities, it soothes to sleep all
nervous hurry and impatience; under its restful influence all the troubles
and vexations and sorrows that harass the mind vanish away, and existence becomes a dream, a charm, a deep and tranquil ecstasy. How it
contrasts with hot and perspiring pedestrianism, and dusty and deafening
railroad rush. . . ."

cruelty. At the same time, however, Clemens resurrects the raft. Near the end of the Grangerford episode (in Chapter 18), Huck discovers that Jim has been hiding in a nearby swamp; when they meet, Jim tells him that he has been working on the raft. Huck is incredulous: " 'You mean to say our old raft warn't smashed all to flinders?' " No, the raft was " 'tore up a good deal — one en' of her was — but dey warn't no great harm done. . . .' " With this simple device, Clemens revives the idyll. The result is like the pulling apart of Ahab's and Ishmael's attitudes in *Moby-Dick;* as Clemens's attack upon society becomes more savage, his treatment of the life identified with the raft becomes more lyrical. In all of our literature, indeed, there is nothing to compare with Huck's incantatory description, at the beginning of Chapter 19, of the sunrise on the river.

> Two or three days and nights went by; I reckon I might say they swum by, they slid along so quiet and smooth and lovely. Here is the way we put in the time. It was a monstrous big river down there — sometimes a mile and a half wide; we run nights, and laid up and hid day-times; soon as night was most gone, we stopped navigating and tied up — nearly always in the dead water under a tow-head; and then cut young cottonwoods and willows and hid the raft with them. Then we set out the lines. Next we slid into the river and had a swim, so as to freshen up and cool off; then we set down on the sandy bottom where the water was about knee deep, and watched the daylight come. Not a sound, anywheres — perfectly still — just like the whole world was asleep, only sometimes the bull-frogs a-cluttering, maybe. The first thing to see, looking away over the water, was a kind of dull line — that was the woods on t'other side — you couldn't make nothing else out; then a pale place in the sky; then more paleness, spreading around; then

the river softened up, away off, and warn't black any more, but gray; you could see little dark spots drifting along, ever so far away — trading scows, and such things; and long black streaks — rafts; sometimes you could hear a sweep screaking; or jumbled up voices, it was so still, and sounds come so far; and by-and-by you could see a streak on the water which you know by the look of the streak that there's a snag there in a swift current which breaks on it and makes that streak look that way; and you see the mist curl up off of the water, and the east reddens up, and the river, and you make out a log cabin in the edge of the woods, away on the bank on t'other side of the river, being a wood-yard, likely, and piled by them cheats so you can throw a dog through it anywheres; then the nice breeze springs up, and comes fanning you from over there, so cool and fresh, and sweet to smell, on account of the woods and the flowers; but sometimes not that way, because they've left dead fish laying around, gars, and such, and they do get pretty rank; and next you've got the full day, and everything smiling in the sun, and the song-birds just going it!

There are countless descriptions of the dawn in literature, yet no substitute exists for this one: it is unique in thought and feeling, in diction, rhythm, and tone of voice. And, above all, it is unique in point of view. Unlike the passengers in "Old Times," or, for that matter, most traditional observers of landscape, this narrator is part of the scene he describes. (At the outset Huck and Jim are sitting in the river with water up to their waists.) He brings to his account no abstract, *a priori* idea of beauty. He has no school of painting in mind, nothing to prove, hence there is room in his account for all the facts. "Here," he says, "is the way we put in our time." He composes the scene as he had experienced it, in a sequence of sense impressions. And since all of his senses are alive, a highlight is thrown upon the minute particulars. The fluid situation adds to the feeling of immediacy. Nothing is fixed, absolute, or perfect. The passage gains immensely in verisimilitude from his repeated approximations: "soon as night was *most* gone," "*nearly always* in the dead water," "a *kind of* dull line," "*sometimes* you could hear," "*but sometimes* not that way." He seems to recall the events at the very instant of perception, and nature, too, is captured in process: "the daylight *come*," "paleness, *spreading* around," "river *softened* up," "mist *curl* up," "east *reddens* up," "breeze *springs* up." Everything is alive, everything is changing; the locus of reality is neither the boy nor the river, neither language nor nature, neither the subject nor the object, but the unending interplay between them.

Then, too, Huck "belongs" to the terrain in the sense that his language is native to it. This fact, probably more than any other, accounts for the astonishing freshness of the writing. Sunrises have not changed much since Homer sang of the rosy-fingered dawn, but here, no doubt, is the first one ever described in this idiom. Its distinctive quality resides in the historical distinctiveness of the narrator, his speech, and the culture from which both derive. But particularly his speech, for that is the raw material of this art, and we delight in the incomparable fitness of subject and language. The lyrical intensity of *Huckleberry Finn* may be ascribed to the voice of the narrator — the boy experiencing the event. Of course, no boy ever spoke such a poetic language, but the illusion that we are hearing the spoken word forms a large part of the total illusion of reality. The words on the page direct our attention to life, not to other works of literature, which is only to say that here art is disguised by originality. In adopting the method of vernacular narration, Clemens had freed himself of certain sterile notions of what a writer should write. He had found an alternative to the rhetoric of both passengers and pilots.

While aiming to convey the beauty of the sunrise, accordingly, Clemens permits Huck to report that "by-and-by you could see a streak on the water which you know by the look of the streak that there's a snag there in a swift current which breaks on it and makes that streak look that way." The boy is endowed with a knowledge of precisely those matters of fact which had seemed to impair the pilot's sense of beauty. He recognizes the menacing principle of nature that spells the destruction of steamboats and men. Mingled with the loveliness of the scene are things not so lovely: murderous snags, wood piled by cheats, and — what could be less poetic? — the smell of dead fish. Huck is neither the innocent traveler nor the initiated pilot. He sees the snags but they do not interfere with his pleasure. In his mind the two rivers are one. His willingness to accept the world as he finds it, without anxiously forcing meanings upon it (his language is lacking in abstractions), lends substance to the magical sense of peace the passage evokes. The sequence follows the diurnal cycle: it begins with the sunrise, and it ends, as night falls, with an evocation of cosmic harmony. At evening the raft is in midstream again, and a strain of music coming over the water does the work of a shepherd's flute:

> Sometimes we'd have that whole river all to ourselves for the longest time. Yonder was the banks and the islands, across the water; and maybe a spark — which was a candle in a cabin window — and sometimes on the water you could see a spark or two — on a raft or a

> scow, you know; and maybe you could hear a fiddle or
> a song coming over from one of them crafts.

All the way down the river the boy had been remarking
the lights on the shore; now they form a continuum with
the stars, and his sense of solidarity with the physical universe acquires the intensity of a religious experience. The
two on the raft face the mystery of the creation with the
equanimity of saints:

> It's lovely to live on a raft. We had the sky, up there,
> all speckled with stars, and we used to lay on our backs
> and look up at them, and discuss about whether they
> was made, or only just happened.

The passion we feel here can only be compared with love.
By adopting the point of view of the boy, for whom the
conflict described in "Old Times" does not exist, Clemens
achieves a unified mode of perception. Now the steamboat, like the railroad at Walden Pond, is incorporated in
the vision of order.

> Once or twice at night we would see a steamboat
> slipping along in the dark, and now and then she would
> belch a whole world of sparks up out of her chimbleys,
> and they would rain down in the river and look awful
> pretty; then she would turn a corner and her lights
> would wink out and her pow-pow shut off and leave
> the river still again; and by-and-by her waves would get
> to us, a long time after she was gone, and joggle the
> raft a bit, and after that you wouldn't hear nothing for
> you couldn't tell how long, except maybe frogs or some-
> thing.

Here, for a moment, Clemens restores the sense of
wholeness the steamboat had smashed. Harmony is
achieved, in the classic Virgilian manner, by withdrawal

to a simpler world. From this vantage, power and complexity are momentarily comprehended by unshielded innocence. But Clemens is clear about the momentariness of this stay against confusion; in fact, the tension is not entirely relieved (the steamboat causes waves that "joggle the raft a bit"), and the incantation is followed immediately by another incursion of the great world, this time in the persons of those grandiloquent frauds, the Duke and the Dauphin. Their attitude toward the occupants of the raft is akin to that of the steamboatmen. "Of course," Huck had said, in describing what happened after the smashing of the raft, ". . . that boat started her engines again ten seconds after she stopped them, for they never cared much for raftsmen. . . ." The Duke and King don't care either, and, in fact, not caring about people is fairly typical of the population of this feuding, lynching, slave-owning society along the banks of the lower Mississippi. Now, in the middle portion of *Huckleberry Finn,* Clemens sets up a rhythmic alternation between idyllic moments on the river and perilous escapades on shore. With each new episode, as the raft moves southward, the contrast between the two worlds becomes more nearly absolute, and the possibility of imagining a comic resolution virtually disappears. In Chapter 31 he again places the raft in jeopardy, and once more confronts the dilemma he had managed to by-pass earlier.[45]

But this time the threat arises in the hero's mind. After the Wilks episode, when Huck discovers that Jim has been betrayed and captured, he pauses to reconsider his situation. He now realizes that the long journey has "all come to nothing," and this thought revives his "conscience." How would the town interpret his behavior? "It would get all around, that Huck Finn helped a nigger to get his freedom. . . ." The conflict that follows is the climax of

the book. At first Huck listens to an inner voice using the
pulpit rhetoric he equates with law-abiding respectability.
Here, the voice says, ". . . was the plain hand of Providence slapping me in the face and letting me know my
wickedness was being watched all the time from up there
in heaven. . . ." Overborne, he decides to inform Miss
Watson of her escaped slave's whereabouts. But then, having written the note, he feels uneasy. He thinks about Jim,
and his memory goes to work:

> And got to thinking over our trip down the river; and
> I see Jim before me, all the time, in the day, and in the
> night-time, sometimes moonlight, sometimes storms, and
> we a floating along, talking, and singing, and laughing.
> But somehow I couldn't seem to strike no places to
> harden me against him, but only the other kind. I'd see
> him standing my watch on top of his'n, stead of calling
> me, so I could go on sleeping; and see him how glad
> he was when I come back out of the fog. . . .

The theme that connects these sharp pictures of Huck's
relation with Jim, setting them against the abstract moral
code he feels obliged to honor, is caring. In caring for each
other he and Jim had formed a bond whose strength is
now put to a test. The pilot of the monstrous steamboat,
on the other hand, had used his power with an arrogant
negligence — a care-lessness — typical of this raw Mississippi world. And here, in the crisis of Chapter 31, the boy
himself is faced with the prospect of exercising power. Before him is a letter capable — or so he thinks — of returning Jim to slavery. The longer he considers the matter,
the more clearly he recognizes what is at stake. "It was a
close place," he says. "I took it up, and held it in my hand.
I was a trembling, because I'd got to decide, forever,
betwixt two things, and I knowed it."

In pressing the trial of his hero to a final choice,

Clemens relinquishes the possibility, once and for all, of a satisfactory comic resolution. Huck's decision to tear up the letter ("'All right, then, I'll *go* to hell'") is justly recognized as an exalted moment in the record of the American consciousness. It joins the pastoral ideal with the revolutionary doctrine of fraternity. But it is a moment achieved at the cost of literary consistency and order. The boy's language, his knowledge that he is deciding "forever, betwixt two things," suggests an unqualified, irrevocable choice — a total repudiation of an oppressive society. Although aiming at a joyous, affirmative outcome of the journey, Clemens has committed his hero to the noble but extravagant ideal of freedom identified with the raft. It is extravagant in the sense that it cannot be reconciled with the facts tacitly acknowledged earlier, at the end of Chapter 16, in the smashing of the raft. It is extravagant because it is, in a word, unattainable. But then the entire metaphoric structure of the fable, the raft floating helplessly toward slave territory, has an essentially tragic import. Hence the somber undertones at the beginning of the final "adventure" — the long, elaborate escapade at the Phelps farm that comprises the last fifth of the book.

> When I got there it was all still and Sunday-like, and hot and sunshiny — the hands was gone to the fields; and there was them kind of faint dronings of bugs and flies in the air that makes it seem so lonesome and like everybody's dead and gone; and if a breeze fans along and quivers the leaves, it makes you feel mournful, because you feel like it's spirits whispering — spirits that's been dead ever so many years — and you always think they're talking about *you*. As a general thing it makes a body wish *he* was dead, too, and done with it all.[46]

By this time Clemens had resigned himself to the abandonment of the raft and the boy's return to civilization.

But the prospect was disheartening, and it is significant that Huck now suffers a return of the dark mood that had seized him at the outset, in the Widow's house. It is as though Clemens, with little, if any, conscious awareness of the convention, somehow had discovered the tragic thread that invariably runs through the fabric of complex pastoralism. The mournful feeling that Huck brings to the "faint dronings" of the insects is appropriate to the novel's final movement, beginning here, away from the virtually Arcadian perfection of life aboard the raft. Throughout the book, to be sure, the boy has exhibited a remarkable sensitivity to the limitations of ordinary human existence. But it was only in Chapter 31, when he realized that the rogues had betrayed Jim, that he had attached a tragic meaning to the voyage itself. "After all this long journey," he had said, ". . . here was it all come to nothing, everything all busted up and ruined. . . ." Nowhere in *Huckleberry Finn* does Clemens come closer to an explicit statement about the outcome of the American idyll. His deepest feelings are registered by Huck, whose mind now turns to thoughts of death. *Et in Arcadia Ego.* And here, paradoxically, Clemens begins the difficult, not to say impossible, task of pulling the novel back on its original course.

Hence the other, ironic meaning of Huck's premonition of death. Shortly after his arrival at the Phelps farm, when he is mistaken for Tom Sawyer, Huck says that it was "like being born again." He might as well have said that it was like being put to death. In order to contrive a resolution compatible with the comic mode, Clemens in effect kills his hero. Huck and Jim are reduced to subordinate rôles. From the time Tom reappears and takes command, the book goes flat. When you read *Huckleberry Finn,* Ernest Hemingway said in 1935, "you must stop where the

Nigger Jim is stolen from the boys. That is the real end. The rest is just cheating." In spite of all the ingenious efforts to justify the final chapters, this judgment remains as sound as ever. "Cheating" may not be the correct word, but the "real end" does coincide with the crisis of Chapter 31. And the reason, which Hemingway did not elaborate, is that the book drew its power from the contradiction resolved in that scene. It is in essence the same conflict of values that had made itself felt, earlier, when the steamboat had smashed the raft. In both episodes Clemens had tacitly acknowledged the tragic import of the journey — a fact that he was unwilling, or unable, to accept, and that the high jinks at the Phelps farm cannot disguise. In the final sentences, however, the book is saved. When Huck is faced with the prospect of being adopted and civilized, the lost theme is recovered. And when he decides to "light out for the Territory ahead of the rest," Clemens seizes an elusive truth. The tension at the center of this book, he now admits, is the very pulse of reality; it may be suppressed, and it may, for a time, be relieved, but eventually it is bound to reappear in a new form.[47]

By 1884, to be sure, the idea of a refuge in the "Territory" was almost as tenuous as the idea of floating downstream on a raft. Six years later the Superintendent of the Census declared that the frontier no longer existed. After *Huckleberry Finn,* meanwhile, Clemens became more fascinated by the dilemma of technological progress. In *A Connecticut Yankee in King Arthur's Court* (1889), as if to test the promise of mechanization, he set an ingenious Yankee mechanic down in sixth-century England. At the time, apparently, most readers interpreted the book as an affirmation of progressive ideas. But sub-

sequent history has compelled us to take the ironic implications of the fable more seriously: for a while, it is true,
the Yankee's program for industrializing Arthurian England seems to be successful; but then a reaction against
his regime sets in, and the end of his experiment in
accelerated progress is gruesome. The Yankee "Boss" and
his loyal followers take refuge behind one of their proud
inventions, an electrified fence. Soon they are surrounded
by a black mass of corpses—a deadly wall of dead men.
As his friend Clarence puts it afterwards, in the "Postscript," they had been caught in a trap of their own making. "If we stayed where we were, our dead would kill
us; if we moved out of our defences, we should no longer
be invincible. We had conquered; in turn we were conquered." In the closing years of his life Clemens became
as bitterly pessimistic as Henry Adams. One of his final
essays, *What Is Man?* (published privately and anony-
mously in 1906), begins with the subtitle: "a. Man the
Machine." [48]

5

During the past twenty years much has been written,
notably by Lionel Trilling, Richard Chase, and R. W. B.
Lewis, about the contradictions said to embody what is
most distinctive in American thought. Although these
critics have been concerned chiefly with literature, their
method is informed by a comprehensive, or potentially
comprehensive, idea of culture. Trilling made the idea
most explicit in his influential essay "Reality in America"
(1940). There, after rebuking V. L. Parrington for using
the image of flowing currents to figure the history of
American ideas (*Main Currents in American Thought*),

Trilling asserts: "A culture is not a flow, nor even a confluence; the form of its existence is struggle, or at least
debate — it is nothing if not a dialectic." What Trilling is
proposing here may be called a dialectical theory of culture. The "very essence" of a culture, he says, resides in
its central conflicts, or contradictions, and its great artists
are likely to be those who contain a large part of the
dialectic within themselves, "their meaning and power
lying in their contradictions." Whatever its shortcomings
as a universal theory, Trilling's definition has proven remarkably useful in the interpretation of American writing
in the nineteenth century — a period when, as he says,
"an unusually large proportion of . . . notable writers
. . . were such repositories of the dialectic of their
times. . . ." [49]

With brilliant effect, Chase later used this dialectical
concept of culture to explain the special character of the
American novel. In *The American Novel and Its Tradition* (1957), he begins with the assertion that the imagination responsible for the best and most characteristic American fiction "has been shaped by the contradictions and not
by the unities and harmonies of our culture." He admits
that this may be true, in a sense, of all literatures of whatever time and place. But the issue is one of degree, and
the American novel "tends to rest in contradictions and
among extreme ranges of experience. When it attempts
to resolve contradictions, it does so in oblique, morally
equivocal ways. As a general rule it does so either in melodramatic actions or in pastoral idyls. . . ." Hence the difference between American and English novels: the tendency in American writing toward abstraction, away from
the specification of social actualities, and away, above all,
from that preoccupation with those subtle relations of

property, class, and status that form the substance of the great Victorian novel. Our writers, instead of being concerned with social verisimilitude, with manners and customs, have fashioned their own kind of melodramatic, Manichean, all-questioning fable, romance, or idyll, in which they carry us, in a bold leap, beyond everyday social experience into an abstract realm of morality and metaphysics.

Nowhere in criticism is there a more persuasive definition of that elusive quality of Americanness in our classic American books. But if we want to know *why* these qualities are peculiar to our literature, Chase's book offers only a cursory answer. The question, to be sure, did not really engage him. He was interested in literary consequences, not extra-literary causes. Though he admitted that the peculiar traits of American writing must be traceable to the special character of the environment, he did not recognize the bearing upon his theory of certain controlling facts of life in nineteenth-century America. Above all, he ignored the unbelievably rapid industrialization of an "underdeveloped" society. Within the lifetime of a single generation, a rustic and in large part wild landscape was transformed into the site of the world's most productive industrial machine. It would be difficult to imagine more profound contradictions of value or meaning than those made manifest by this circumstance. Its influence upon our literature is suggested by the recurrent image of the machine's sudden entrance into the landscape.

By the turn of the century the device was becoming a cliché of American writing. In *The Octopus* (1901), Frank Norris used it to provide a dramatic, premonitory climax for the opening chapter. Presley, a poet, is walking at sunset in the rich San Joaquin Valley. It is a lovely, mild evening. Everything is very still. He is in a revery. "All about, the feeling of absolute peace and quiet and security and untroubled happiness and content seemed descending from the stars like a benediction. The beauty of his poem, its idyl, came to him like a caress. . . . But suddenly there was an interruption." By now it is dark, and the train, "its enormous eye, Cyclopean, red, throwing a glare far in advance," comes thundering down the track and smashes into a herd of sheep.

> The iron monster had charged full into the midst, merciless, inexorable. To the right and left, all the width of the right of way, the little bodies had been flung; backs were snapped against the fence posts; brains knocked out. Caught in the barbs of the wire, wedged in, the bodies hung suspended. Under foot it was terrible. The black blood, winking in the starlight, seeped down into the clinkers between the ties with a prolonged sucking murmur.

Sick with horror, Presley turns away. "The sweetness was gone from the evening, the sense of peace . . . stricken from the landscape." Walking on, he hears the whistle of the engine, faint and prolonged, reverberating across the sweep of land. He again sees the machine in his imagination, a "terror of steel and steam," hurtling across the horizon; but now he sees it as "the symbol of a vast power, huge, terrible, flinging the echo of its thunder over all the reaches of the valley . . . ; the leviathan, with tentacles of steel clutching into the soil, the soulless Force, the iron-hearted Power, the monster, the Colossus, the Octopus." [50]

There is no point in piling up examples. The omnipresence of the device is significant chiefly because it relates the special conditions of life in America with that tendency of mind to which Richard Chase directed attention: a habit of defining reality as a contradiction between radically opposed forces. No single work makes this connection more explicit or dramatic than *The Education of Henry Adams* (privately printed in 1907). A major theme of the book is the pulling apart, in Adams's own experience as in the culture generally, of feeling and intellect, love and power. Adams adroitly builds the story up to 1900 and his discovery, in a moment of religious exaltation, of the relation between his two master symbols — the Dynamo and the Virgin. But in order to set his theme, in the first chapter, Adams turns back to the year of the "little event" recorded by Hawthorne in Sleepy Hollow. It was in 1844, as he sees it in retrospect, that the forces of history bore down upon him and fixed his fate. That was the year, he says, when

> . . . the old universe was thrown into the ash-heap and a new one created. He and his eighteenth-century, troglodytic Boston were suddenly cut apart — separated forever — in act if not in sentiment, by the opening of the Boston and Albany Railroad; the appearance of the first Cunard steamers in the bay; and the telegraphic messages which carried from Baltimore to Washington the news that Henry Clay and James K. Polk were nominated for the Presidency. This was in May, 1844; he was six years old; his new world was ready for use, and only fragments of the old met his eyes.

A sense of the transformation of life by technology dominates *The Education* as it does no other book. This is partly because Adams comes at the theme with the combined techniques of the cool historian and the impassioned poet. In the excerpt above, as in *The Education* generally, we can distinguish two voices. The voice of the historian tells us the news. He regards industrial power as an objective "cause" of change in American society. And he is able to draw a direct line of connection from these changes to his own life. It is technology (the new railroads, steamboats, and telegraph) that has *separated* Adams from his family's eighteenth-century tradition. Writing in this vein Adams virtually endorses a theory of technological determinism. "As I understand it," he wrote to his brother while at work on the book, "the whole social, political and economical problem is the resultant of the mechanical development of power." Throughout *The Education* we hear the voice of the historian calling attention to technology as an impersonal and largely uncontrolled force acting upon human events. The historian provides us with statistics on accelerating coal and steel production; he makes us aware of the impact of the new power upon society quite apart from the way it strikes him. But the voice of the poet continually joins in. He charges historical fact with emotion by using such images as "thrown into the ash-heap," and submerged metaphors: "suddenly cut apart — separated forever." The poet invests the statement with traumatic overtones, so that the sudden appearance of the machine has the effect of irrevocable separation like the cutting of an umbilical cord. The historian tells us what happened to America and to Adams, but the poet tells us how we should feel about what happened.[51]

Moving back and forth between conceptual statements about the growth of power and sensory impressions of the same process, Adams raises his theme to a melodramatic pitch. Machine imagery serves to convey both an inward and a historical contradiction. Early in the book he describes a journey he took in 1858 through the English "Black District" [sic]. It was, he writes, like a "plunge into darkness lurid with flames." There, in the center of industrial England, Adams suddenly had a "sense of unknown horror in . . . [the] weird gloom which then existed nowhere else, and never had existed before, except

in volcanic craters." Above all, he stresses an acute sense
of the "violent contrast between this dense, smoky, impenetrable darkness, and the soft green charm that one
glided into, as one emerged. . . ." The scene is England,
but the sensibility is American. What strikes the American
with particular force is the violence of the contrast between the industrial and the natural landscapes. In the
work of Carlyle or Ruskin or Morris we may find similar
passages, but in general the English writer is more likely
to regard the new power as a threat to some cherished
ideal of high civilization or art or craftsmanship.[52]

Adams's evocation of horror in the presence of the machine reaches its climax in the account of the Paris Exposi-
tion of 1900 where "he found himself lying in the Gallery
of Machines . . . , his historical neck broken by the sudden
irruption of forces totally new." Confronting the new
dynamo he felt an impulse to pray to it, much as Chris-
tians prayed to the cross — or to the Virgin. Here all the
polarities of the book, among them the violent contrast
between the weird, smoky gloom and the soft green charm
of the landscape, culminate in a symbolic tableau depicting "two kingdoms of force which had nothing in common
but attraction": one represented by the Dynamo, the
other by the Virgin. If the tendency toward an abstract
or dialectical view of life is a distinctive characteristic of
American culture, then *The Education of Henry Adams*
is one of the most American of books. Adams uses the opposition between the Virgin and the Dynamo to figure an
all-embracing conflict: a clash between past and present,
unity and diversity, love and power. In his Manichean
fashion he marshals all conceivable values. On one side
he lines up heaven, beauty, religion, and reproduction; on
the other: hell, utility, science, and production.[53]

Although devoid of pastoral feelings, this melodramatic

tableau is a characteristic product of complex, twentieth-century American pastoralism. It embodies a view of life that may be traced to the Sleepy Hollow notes of 1844. To be sure, there is a world of difference between the highly wrought pattern contrived by Adams and the sketchy impression of the new machine power set down by Hawthorne. What had begun as a casual notation now appears as an elaborate, tragic, and all-inclusive thematic figure. And yet it is impossible to miss the continuity between the two. In the first chapter of *The Education* Adams introduces a variant of the pastoral design in describing the rhythmic alternation between the seasonal attitudes toward nature and civilization he had known as a boy:

> Winter and summer, then, were two hostile lives, and bred two separate natures. Winter was always the effort to live; summer was tropical license. Whether the children rolled in the grass, or waded in the brook, or swam in the salt ocean, or sailed in the bay, or fished for smelts in the creeks, or netted minnows in the salt-marshes, or took to the pine-woods and the granite quarries, or chased muskrats and hunted snapping-turtles in the swamps, or mushrooms or nuts on the autumn hills, summer and country were always sensual living, while winter was always compulsory learning. Summer was the multiplicity of nature; winter was school.

By invoking the cycle of the seasons, Adams seems to point toward an orthodox pastoral resolution. Yet even here, at the outset, he stresses the ineradicable character of this root conflict. The passage continues:

> The bearing of the two seasons on the education of Henry Adams was no fancy; it was the most decisive force he ever knew; it ran through life, and made the

division between its perplexing, warring, irreconcilable
problems, irreducible opposites, with growing emphasis
to the last year of study. From earliest childhood the
boy was accustomed to feel that, for him, life was double.
Winter and summer, town and country, law and liberty,
were hostile, and the man who pretended they were not,
was in his eyes a schoolmaster — that is, a man employed
to tell lies to little boys.

The apocalyptic image of the Dynamo and the Virgin
is the ultimate expression of the tragic doubleness that
Adams locates at the center of modern history. To be sure,
the encounter with the machine is a portentous event in
earlier works of the American imagination. In *Walden*
and *Moby-Dick* and *Huckleberry Finn* it creates awe and
terror and a sense of powerlessness. And these fables do
tend to "rest," as Richard Chase put it, "in contradictions
and among extreme ranges of experience." Nevertheless,
Adams imparts a new sense of violence, inclusiveness, and
irreconcilability to the clash between the two kingdoms
of force. His Dynamo is the modern force most nearly
equivalent, in its command over human behavior, to the
sublimated sexual vitality that built Chartres. But in all
other respects the two forces are radically, irrevocably
opposed. One resists, the other exalts, the dominion of
Eros. In the symbol of the Dynamo, Adams represents an
industrial society that threatens, as no previous society
has threatened, to destroy the creative power embodied
in the Virgin. To him she represents "the highest energy
ever known to man, the creator of four-fifths of his noblest
art, exercising vastly more attraction over the human mind
than all the steam-engines and dynamos ever dreamed
of. . . ."

In *The Education of Henry Adams*, then, the onset of

machine power initiates a fatal chain of events. The forces that came between young Adams and his "old universe" in 1844, suddenly cutting them apart, separating them forever, were to produce the Dynamo and a nightmare vision of race suicide. As early as 1862, in fact, Adams had told his brother Charles:

> You may think all this nonsense, but I tell you these are great times. Man has mounted science and is now run away with. I firmly believe that before many centuries more, science will be the master of man. The engines he will have invented will be beyond his strength to control. Some day science may have the existence of mankind in its power, and the human race commit suicide by blowing up the world.

To use the Adams idiom, the sound that Hawthorne had heard in Sleepy Hollow in 1844 was a new, perhaps final, outbreak of the ancient war between the kingdom of love and the kingdom of power, and it has been waged endlessly in American writing ever since.[54]

In 1904, after a long absence in Europe, Henry James was impressed once again by the strangely persistent affinity between the American landscape and the pastoral ideal. Having traveled to New Hampshire almost immediately after landing in New York, the first thing that strikes him (as he later explains in *The American Scene,* published in 1907) is the Arcadian aspect of his native land. He awakes in a place full of "the sweetness of belated recognition, that of the sense of some bedimmed summer of the distant prime flushing back into life and asking to give again as much as possible of what it had given before — all in spite, too . . . of the newness, to my eyes . . . of the particular rich region." Admitting at once that there is something odd about using the word "rich" to describe a vista of such manifest "poverties," James says that nevertheless he uses the word without compunction, "caring little that half the charm, or half the response to it, may have been shamelessly 'subjective' . . ." The light and playful tone dismisses (or attempts to dismiss) the problem that had tried the generation of Emerson, Thoreau, and Melville. Is James attributing to visible nature qualities which do not belong there? He says that he does not care whether they are there or not, "since that but slightly shifts the ground . . . of the impression." Perhaps the pastoral aspect is in the eye of the beholder, but what if it is? "When you wander about in Arcadia," he explains, slyly dodging the issue, "you ask as few questions as possible. That *is* Arcadia in fact, and questions drop, or at least get themselves deferred and shiftlessly shirked. . . ." [55]

But he is too sanguine; the question will not drop. Alas, there is something "curious, unexpected, inscrutable" about the effect of this landscape. "Why," he asks, is the "whole connotation so *delicately* Arcadian, like that of the Arcadia of an old tapestry, an old legend, an old love-story in fifteen volumes . . . ?" He cannot shirk the question after all, and he rephrases it: why, in the absence of other elements of the "higher finish," do these American views "insist on referring themselves to the idyllic *type* in its purity?" What puzzles him, above all, is that a landscape so raw and new should evoke connotations so refined, archaic, and, in a word, typical. It is as if, he says, "the higher finish, even at the hand of nature, were in some sort a perversion," and that these fascinating landscapes seem "more exquisitely and ideally Sicilian, Theocritan, poetic, romantic, academic, from their not bearing the burden of too much history." Precisely because it is relatively unformed, wild, and new, James is saying, the scenery of America is peculiarly hospitable to pastoral illusions. It invites us to cross the commonsense boundary between art and reality, to impose literary ideas upon the world.

How misleading this pastoral bent can be does not become evident until James explores the region near Farmington, Connecticut. Once again, the village scene renews that ancient vision "of the social idyll, of the workable, the expensively workable, American form of country life; and, in especial, of a perfect consistency of surrender to the argument of the verdurous vista." Here, says James, is the American village at its best, where a "great elm-gallery happens to be garnished with old houses, and the old houses happen to show style and form and proportion. . . ." What finally is most striking, however, about the picture of the white village with its high thin church steeple, so archaic in modern America as to seem almost a heraldic emblem, is another object close by. It is the railway crossing. Out of the contrast James gains a new perception. While the church now seems a mere monument embellishing some large white card, the railroad becomes for him the "localization of possible death and destruction," and, set against the rural scene as it is, the industrial image evokes an over-all impression of a kind of "monotony of acquiescence." By the look of "universal acquiescence" he means, among other things, that there is nothing in the scene capable of resisting the domination of the machine. The "common man" and "common woman," he says, with a superior tone, are defenseless. "The bullying railway orders them off their own decent avenue without a fear that they will 'stand up' to it. . . ." There is nothing in the visible landscape — no tradition, no standard, no institution — capable of standing up to the forces of which the railroad is the symbol. As the work of Thoreau and Melville and Mark Twain had testified, the pastoral dream was no defense. "This look as of universal acquiescence," says James, "plays somehow through the visible vacancy — seems a part of the thinness, the passivity, of that absence of the settled standard which contains . . . the germ of the most final of all my generalizations." [56]

What James sees in Farmington, then, is not one among many American scenes, but a thematic or symbolic landscape. It figures a controlling idea, the very marrow of his subject, which happens to be America. The contrast between the machine and the pastoral ideal dramatizes the great issue of our culture. It is the germ, as James puts it, of the most final of all generalizations about America.

Epilogue: The Garden of Ashes

In various quiet nooks and corners I had the beginnings of
all sorts of industries under way — nuclei of future vast factories, the iron and steel missionaries of my future civilization.
In these were gathered together the brightest young minds
I could find, and I kept agents out raking the country for
more, all the time. I was training a crowd of ignorant folk
into experts — experts in every sort of handiwork and scientific calling. These nurseries of mine went smoothly and privately along undisturbed in their obscure country retreats.

.   .   .   .

My works showed what a despot could do with the resources of a kingdom at his command. Unsuspected by this
dark land, I had the civilization of the nineteenth century
booming under its very nose! It was fenced away from the
public view, but there it was, a gigantic and unassailable fact
— and to be heard from, yet, if I lived and had luck. There
it was, as sure a fact and as substantial a fact as any serene
volcano, standing innocent with its smokeless summit in the
blue sky and giving no sign of the rising hell in its bowels.

Mark Twain, *A Connecticut Yankee in
King Arthur's Court*, 1889

"So," say the parable-makers, "is your pastoral
life whirled past and away." We cannot deny the fact
without denying our history. When the Republic was
founded, nine out of ten Americans were husbandmen;
today not one in ten lives on a farm. Ours is an intricately
organized, urban, industrial, nuclear-armed society. For
more than a century our most gifted writers have dwelt
upon the contradiction between rural myth and technological fact. In the machinery of our collective existence,
Thoreau says, we have "constructed a fate, an *Atropos*"

that never will turn aside. And until we confront the unalterable, he would add, there can be no redemption from a system that makes men the tools of their tools. A similar insight informs *Moby-Dick*. But in the penchant for illusion Melville saw more dire implications. It is suicidal, Ishmael learns, to live "as in a musky meadow" when in truth one is aboard a vessel plunging into darkness. What was a grim possibility for Melville became a certainty for Mark Twain and Henry Adams; neither was able to imagine a satisfactory resolution of the conflict figured by the machine's incursion into the garden. By the turn of the century they both envisaged the outcome as a vast explosion of new power. Power, Adams said, now leaped from every atom. The closing chapters of *The Education of Henry Adams* are filled with images of mankind in the grip of uncontrollable forces. He pictured the forces grasping the wrists of man and flinging him about "as though he had hold of . . . a runaway automobile. . . ." Adams was haunted by the notion that bombs were about to explode. "So long as the rates of progress held good," he observed, "these bombs would double in force and number every ten years." [1]

But the ancient ideal still seizes the native imagination. Even those Americans who acknowledge the facts and understand the fables seem to cling, after their fashion, to the pastoral hope. This curious state of mind is pictured by Charles Sheeler in his "American Landscape" (1930; plate 3). Here at first sight is an almost photographic image of our world as it is, or, rather, as we imagine it will be if we proceed without a change of direction. No trace of untouched nature remains. Not a tree or a blade of grass is in view. The water is enclosed by man-made banks, and the sky is filling with smoke. Like the reflection upon the water, every natural object represents some aspect of the collective economic enterprise. Technological power overwhelms the solitary man; the landscape convention calls for his presence to provide scale, but here the traditional figure acquires new meaning: in this mechanized environment he seems forlorn and powerless. And yet, somehow, this bleak vista conveys a strangely soft, tender feeling. On closer inspection, we observe that Sheeler has eliminated all evidence of the frenzied movement and clamor we associate with the industrial scene. The silence is awesome. The function of the ladder (an archaic implement) is not revealed; it points nowhere. Only the minuscule human figure, the smoke, and the slight ripples on the water suggest motion. And the very faintness of these signs of life intensifies the eerie, static, surrealist quality of the painting. This "American Landscape" is the industrial landscape pastoralized. By superimposing order, peace, and harmony upon our modern chaos, Sheeler represents the anomalous blend of illusion and reality in the American consciousness.

In F. Scott Fitzgerald's *The Great Gatsby* (1925), another image of landscape provides the indispensable clue to the baffling career of an archetypal American. Like the other major characters, James Gatz is a Westerner who has come East to make his fortune. (In this version of the fable, as in the "international" novels of Henry James, the direction of the journey is reversed: Gatsby moves from simplicity to sophistication.) In the end, Fitzgerald's narrator, Nick Carraway, realizes that Gatsby can be understood only in relation to the "last and greatest of all human dreams": the original European vision of the "fresh, green breast of the new world." It is Nick who must decide, finally, upon the value of that dream. His problem is akin to the one Shakespeare's Prospero had faced three centuries before; to make sense of his experience Nick must define the relative validity of the "garden" and the "hideous wilderness" images of the New World; he must discriminate between Gatsby's image of felicity, represented by Daisy Buchanan and the "green light," and the industrial landscape of twentieth-century America.

From the beginning Nick is aware of something odd about the elegant green lawns of suburban Long Island. They are green enough, but somehow synthetic and delusive. "The lawn," says Nick, describing his first impression of the Buchanan's home, "started at the beach and ran toward the front door for a quarter of a mile, jumping over sun-dials and brick walks and burning gardens — finally when it reached the house drifting up the side in bright vines as though from the momentum of its run." And then, in the next sentence, he introduces Tom Buchanan: hard, supercilious, arrogant. "Not even the effeminate swank of his riding clothes," says Nick, "could hide the enormous power of that body. . . ." Nick soon discovers that suburban greenness, like Tom's clothing, is misleading. It too is a mask of power, and on the way to Manhattan he sees a more truly representative landscape of this rich and powerful society:

> About half way between West Egg and New York the motor road hastily joins the railroad and runs beside it for a quarter of a mile, so as to shrink away from a certain desolate area of land. This is a valley of ashes — a fantastic farm where ashes grow like wheat into ridges and hills and grotesque gardens; where ashes take the forms of houses and chimneys and rising smoke and, finally, with a transcendent effort, of men who move

dimly and already crumbling through the powdery air. Occasionally a line of gray cars crawls along an invisible track, gives out a ghastly creak, and comes to rest, and immediately the ash-gray men swarm up with leaden spades and stir up an impenetrable cloud, which screens their obscure operations from your sight.[2]

This hideous, man-made wilderness is a product of the technological power that also makes possible Gatsby's wealth, his parties, his car. None of his possessions sums up the quality of life to which he aspires as well as the car. "It was a rich cream color, bright with nickel, swollen here and there in its monstrous length with triumphant hat-boxes and supper-boxes and tool-boxes, and terraced with a labyrinth of wind-shields that mirrored a dozen suns." Sitting behind these many layers of glass, says Nick, was like being "in a sort of green leather conservatory." As it happens, the car proves to be a murder weapon and the instrument of Gatsby's undoing. The car and the garden of ashes belong to a world, like Ahab's, where natural objects are of no value in themselves. Here all of visible nature is as expendable as a pasteboard mask:

> Every Friday five crates of oranges and lemons arrived from a fruiterer in New York — every Monday these same oranges and lemons left his back door in a pyramid of pulpless halves. There was a machine in the kitchen which could extract the juice of two hundred oranges in half an hour if a little button was pressed two hundred times by a butler's thumb.

In *The Great Gatsby,* as in *Walden, Moby-Dick,* and *Huckleberry Finn,* the machine represents the forces working against the dream of pastoral fulfillment.[3]

But in Fitzgerald's book, as in many American fables, there is another turn to the story. Somehow, in spite of the counterforce, the old dream retains its power to stir the imagination. From the first moment Nick sees Gatsby, standing alone in the moonlight gazing across the water at "nothing except a single green light," he is aware of a mysterious, indefinable, transcendent quality in the man. At the outset he describes this appealing trait in deliberately vague language: ". . . there was," he says, "something gorgeous about him, some heightened sensitivity to the promises of life . . . an extraordinary gift for hope, a romantic readiness such as I have never found in any other person and which it is not likely I shall ever find again." Nick's job, as narrator, is to specify the exact nature of the "something" — to find out what it is and where it arises. But the more he learns about Gatsby the more difficult the job becomes. He can find no way to reconcile the man's engaging qualities with the facts: the interminable, senseless parties in Gatsby's "blue gardens," his vulgar show of wealth, his bogus Norman mansion, his monstrous car, and, above all, his bland complicity in crime. What mystifies Nick is the incongruity of the sordid facts, combined as they are with Gatsby's unswerving devotion to Daisy and the ideal of the green light. Only at the very end, when Nick returns to Long Island for the last time, does he find a wholly plausible explanation. And it is here, in the closing paragraphs, that Fitzgerald enlarges the meaning of his fable, extending it to an entire culture that is peculiarly susceptible, like Jay Gatsby, to pastoral illusions. One night, after Gatsby's death, Nick comes back to Long Island. The summer is over. What he now sees, as he sprawls on the beach looking across the water, enables him to place the events he has witnessed in the context of history:

> Most of the big shore places were closed now and there were hardly any lights except the shadowy, moving glow

of a ferryboat across the Sound. And as the moon rose higher the inessential houses began to melt away until gradually I became aware of the old island here that flowered once for Dutch sailors' eyes — a fresh, green breast of the new world. Its vanished trees, the trees that had made way for Gatsby's house, had once pandered in whispers to the last and greatest of all human dreams; for a transitory enchanted moment man must have held his breath in the presence of this continent, compelled into an aesthetic contemplation he neither understood nor desired, face to face for the last time in history with something commensurate to his capacity for wonder.

Here, for the first time, Nick locates the origin of that strange compound of sentiment and criminal aggressiveness in Gatsby. Although this vision of the American landscape reverses the temporal scheme of Sheeler's painting (instead of imposing the abstract residuum of the pastoral dream upon the industrial world, Nick's vision discloses the past by melting away the inessential present) — it serves much the same purpose. It also represents the curious state of the modern American consciousness. It reveals that Gatsby's uncommon "gift for hope" was born in that transitory, enchanted moment when Europeans first came into the presence of the "fresh, green breast of the new world." We are reminded of Shakespeare's Gonzalo and Miranda, of Robert Beverley and Crèvecœur and Jefferson: in America hopefulness had been incorporated in a style of life, a culture, a national character. Hence Gatsby's simple-minded notion that everything can be made right again. Daisy is for him what the green island once had been for Dutch sailors; like them he mistakes a temporary feeling for a lasting possibility.

As Fitzgerald's narrator pieces it together, accordingly, Gatsby's tragic career exemplifies the attenuation of the pastoral ideal in America. In the beginning Nick compares Gatsby's "heightened sensitivity to the promises of life" to a seismograph — a delicate instrument peculiarly responsive to invisible signals emanating from the land. Gatsby's entire existence — not only the "romantic readiness" of his spirit, but also his Horatio Alger rise to affluence — had been shaped by the special conditions of which the bountiful, green landscape is the token. Young James Gatz got his start by using his intimate knowledge of the Midwestern terrain to save a rich man's yacht. It was at that moment that Jay Gatsby sprang, like a son of God, from his Platonic conception of himself. The incident, a turning point in his life, marks the enlistment of native energies in the service of wealth, status, power, and, as Nick puts it, of "a vast, vulgar, and meretricious beauty." Nick, the real hero of *The Great Gatsby*, is the only one, finally, to understand, but it takes him a long while to grasp the subtle interplay between Gatsby's dream and his underworld life.[4]

In part Nick's difficulty arises from the fact that he is so much like Gatsby. He too is a Westerner, a simple man in a complex society, and when he pays his first visit to the Buchanans, he admits, their acquired Eastern manners make him uncomfortable. He feels like a rustic. " 'You make me feel uncivilized, Daisy,' " he said. " 'Can't you talk about crops or something?' " And then, just before describing the sordid, drunken brawl in Mrs. Wilson's apartment, Nick makes explicit his own propensity to Virgilian fantasies:

> We drove over to Fifth Avenue, so warm and soft, almost pastoral, on the summer Sunday afternoon that

I wouldn't have been surprised to see a great flock of white sheep turn the corner.

The party, as Nick describes it, is a debased bucolic festival: the pathetic little dog at the center, the furniture tapestried with "scenes of ladies swinging in the gardens of Versailles," the brutal outburst of Buchanan's hate, and Nick's yearning, through it all, to get away: "I wanted to get out," he says, "and walk eastward toward the Park through the soft twilight. . . ." Because he, too, is drawn to images of pastoral felicity, Nick is prepared to recognize the connection between Gatsby's sentimentality and the sight that had greeted Dutch sailors three centuries before.

> And as I sat there brooding on the old, unknown world, I thought of Gatsby's wonder when he first picked out the green light at the end of Daisy's dock. He had come a long way to this blue lawn, and his dream must have seemed so close that he could hardly fail to grasp it.

And here, at last, Nick pulls back, separating himself from his dead friend's dead dream.

> He [Gatsby] did not know that it was already behind him, somewhere back in that vast obscurity beyond the city, where the dark fields of the republic rolled on under the night.[5]

The difference between Gatsby's point of view and Nick's illustrates the distinction, with which I began, between sentimental and complex pastoralism. Fitzgerald, through Nick, expresses a point of view typical of a great many twentieth-century American writers. The work of Faulkner, Frost, Hemingway and West comes to mind. Again and again they invoke the image of a green landscape — a terrain either wild or, if cultivated, rural — as a symbolic repository of meaning and value. But at the same time they acknowledge the power of a counter-force, a machine or some other symbol of the forces which have stripped the old ideal of most, if not all, of its meaning. Complex pastoralism, to put it another way, acknowledges the reality of history. One of Nick Carraway's great moments of illumination occurs when he realizes that Gatsby wants nothing less of Daisy than that she should go to Tom and say, " 'I never loved you.' " And when Nick objects, observing that one cannot undo the past, Gatsby is incredulous. Of course he can. " 'I'm going to fix everything,' " he says, " 'just the way it was before. . . .' " Like Melville's Starbuck, Gatsby would let faith oust fact. He is another example of the modern primitive described by Ortega, the industrial *Naturmensch* who is blind to the complexity of modern civilization; he wants his automobile, enjoys it, yet regards it as "the spontaneous fruit of an Edenic tree." Nick also is drawn to images of pastoral felicity, but he learns how destructive they are when cherished in lieu of reality. He realizes that Gatsby is destroyed by his inability to distinguish between dreams and facts. In the characteristic pattern of complex pastoralism, the fantasy of pleasure is checked by the facts of history.[6]

But what, then, do our fabulists offer in place of the ideal landscape? Nick Carraway's most significant response to Gatsby's death is his decision, announced in the opening pages of the book, to return to the West. (Like Thoreau, Melville and Mark Twain, Fitzgerald tells his tale from the viewpoint of a traveler returned from a voyage of initiation.) In the closing pages Nick again explains that the East was haunted for him after the murder; and so, he says, "when the blue smoke of brittle leaves was in the air . . . I decided to come back home." Nick's repudiation of the East is a belated, ritualistic withdrawal in the direction of "nature." It is ironically set against the fact, which the entire novel makes plain, that the old distinction between East and West has all but disappeared. Nick's final gesture is a mere salute to the memory of a vanished America; and the book ends on a sadly enervated note of romantic irony: "So," Nick says, "we beat on, boats against the current, borne back ceaselessly into the past."

The ending of *The Great Gatsby* reminds us that American writers seldom, if ever, have designed satisfactory resolutions for their pastoral fables. The power of these fables to move us derives from the magnitude of the protean conflict figured by the machine's increasing domination of the visible world. This recurrent metaphor of contradiction makes vivid, as no other figure does, the bearing of public events upon private lives. It discloses that our inherited symbols of order and beauty have been divested of meaning. It compels us to recognize that the aspirations once represented by the symbol of an ideal landscape have not, and probably cannot, be embodied in our traditional institutions. It means that an inspiriting vision of a humane community has been reduced to a token of individual survival. The outcome of *Walden, Moby-Dick,* and *Huckleberry Finn* is repeated in the typical modern version of the fable; in the end the American hero is either dead or totally alienated from society, alone and powerless, like the evicted shepherd of Virgil's eclogue. And if, at the same time, he pays a tribute to the image of a green landscape, it is likely to be ironic and bitter. The resolutions of our pastoral fables are unsatisfactory because the old symbol of reconciliation is obsolete. But the inability of our writers to create a surrogate for the ideal of the middle landscape can hardly be accounted artistic failure. By incorporating in their work the root conflict of our culture, they have clarified our situation. They have served us well. To change the situation we require new symbols of possibility, and although the creation of those symbols is in some measure the responsibility of artists, it is in greater measure the responsibility of society. The machine's sudden entrance into the garden presents a problem that ultimately belongs not to art but to politics.

Afterword

The Machine in the Garden began as a doctoral dissertation I wrote in 1949, some fifteen years before it became a book.[1] It was an exploratory study of literary responses to the onset of industrialism in America. The subject had caught my attention a decade earlier, when I was an undergraduate studying history and literature in the charged political atmosphere of the late 1930s.

In 1937, when my class entered college, the crisis of the industrialized world was acute. The American economy had not recovered from the financial collapse and mass unemployment of the Great Depression, and it was doubtful whether capitalism could survive another economic downturn. In Europe, meanwhile, the democracies were threatened by the military buildup of Nazi Germany and the Soviet Union. Then, in 1939, in quick succession, came the fascist victory in the Spanish Civil War, the Nazi-Soviet pact, and the outbreak of the Second World War. By the time my class graduated, in June, 1941, the military draft was in effect, and most of us soon were wearing military uniforms.

During our four years as undergraduates, a sizeable minority of the Harvard class of 1941 became leftists. The near collapse of the capitalist system, coupled with the apparent vigor of Soviet socialism, had given unprecedented credence to the ideology of the Left. At that time, before Marxism had been tainted by association with Stalinism, many of us identified left-wing radicalism with the heroic defense of Madrid, last bastion of a democracy besieged by the fascist armed forces of Franco, Mussolini, and Hitler. To

us the Spanish cause epitomized the spirit of Left interna-
tionalism. It later was painful to discover how ill-informed
and naïve we had been about the Stalinist regime, but the
concept of the totalitarian state as a distinctive political form
had not yet been formulated, and the differences between
Nazism and Stalinism seemed more important than the similarities. As an undergraduate I belonged to the Harvard
Student Union, a branch of the American Student Union,
which later won a place on the U.S. Attorney General's infa-
mous list of "subversive organizations." We supported local
labor unions, organized demonstrations against war and fas-
cism, sponsored a lively theatre group, and published our
own magazine of opinion, *The Harvard Progressive* (modeled
on *The Nation* and *New Masses*) of which I was editor. The
Union raised money for the Spanish cause, and we persist-
ently tried—and failed—to devise an alternative to the official American foreign policy that might reconcile our
antiwar and antifascist inclinations.

Many of our teachers also were partisans of the Left during those years. At my first meeting with Daniel Boorstin,
my sophomore-year tutor, he felt obliged to warn me that
he was a member of the Communist Party. Both F. O.
Matthiessen and Paul Sweezy, with whom I would have lasting friendships, were outspoken socialists, and most of the
other faculty members whom I knew well—Harry Levin,
Ralph Barton Perry, Perry Miller—were staunch liberals of
a left–New Deal persuasion. The academic respect then
accorded to anti-capitalist thought was exemplified, in my
formal education, by a memorable course—*Economics 11:
The Economics of Socialism*—taught jointly by Paul Sweezy, a
prominent Marxist economist, and Edward S. Mason, a
Keynesian. In alternating lectures they expounded
opposed views of each topic in the syllabus—a procedure
that was unusually successful in provoking students to form

their own judgments about complex real-world issues. (The Vietnam era was the only time, in my academic experience, that faculty members gave as much attention to the contested issues of the day.) The driving force of history, in Sweezy's understanding of historical materialism, was the growth of human productivity and—by extension—of capital, especially in the form of innovative technologies.[2]

Soon after graduating in 1941, I began a four-year stint in the U.S. Navy. My wartime experience reinforced some of the views I had acquired in college. Late in the Pacific war, for example, my ship was assigned to patrol duty off the entrance to Pearl Harbor, where we had a close-up look at a parade of the sleek, state-of-the-art warships coming out of American shipyards. It was a persuasive demonstration of the critical role of newly developed technologies in building national power. Then, shortly before I left the Navy in October, 1945, came Hiroshima. Quite apart from its tragic human consequences, no other event in my lifetime so effectively dramatized the nexus between science-based technological progress and the cumulative, long-term degradation of the environment. Having been an enthusiastic wilderness camper and amateur ornithologist, I had already become sensitive to our society's increasingly reckless assault on the integrity of the natural environment.

Of the attitudes I had formed as an undergraduate, at least two—one political and the other literary—shaped *The Machine in the Garden.* For some time after the war I continued to think of myself as a socialist. Though that commitment eventually lost most of its practicability, it informed my sense of political history and, more particularly, my belief that industrialization—the capitalist-driven process by which a predominantly rural and agricultural society became predominantly urban and industrial—was the most important "event" in American history. As an undergraduate working with Perry Miller and F. O. Matthiessen, my chief interest had been the interplay between imaginative literature and social change; another lasting attitude, then, was my conviction that some of the most sensitive, telling responses to complex changes in society are to be found in the work of accomplished artists and writers. Though poetry and fiction are not very helpful in establishing the historical record as such, they are singularly useful, I learned, in getting at the more elusive, intangible effects of change—its impact on the moral and aesthetic, emotional and sensory, aspects of experience. Like many intellectuals on the Left in the 1930s, however, I found it difficult to reconcile my political and aesthetic preferences.

Early in my canvass of materials for the dissertation, I considered writing about the "proletarian" left-wing novels of the 1930s. No other body of American writing contains as forthright and critical an account of working-class life in the regime of industrial capitalism. Implicit in most of those books is a principled skepticism, like my own, about the ultimate compatibility of capitalism with the goal of creating a peaceful, just society. The more I read in these earnest, committed novels, however, the more predictable and formulaic, the less interesting, they seemed. Then I came upon a book review by the eminent critic, Edmund Wilson, in which he rebuked the author, Maxwell Geismer, for implying that the first serious response to industrialism in American literature was to be found in the protest novels of the 1930s.[3] Wilson argued that, on the contrary, the real groundbreaking had been accomplished by the writers of the 1880s and 1890s. Since I then happened to be immersed in the literature of the 1830–1860 era, the erudite Wilson's invocation of the later date astonished me. Granted that the earlier work was not in any explicit sense "about" industrialization, it nonetheless was filled with striking sensory images of the artifacts of industrial capitalism: canals, steam engines, steamboats, locomotives, the telegraph, looms, and factories. I never discovered whether Wilson really was ignorant of this salient fact, but his uninformed comment encouraged me to suppose that I had hit upon a subject worth pursuing.

I decided to look closely at the responses of several accomplished writers to that primordial (early nineteenth-century) phase of industrialization. They proved to be fascinating, bearing out Ezra Pound's aphorism, "Artists are the antennae of the race." That the larger significance of these early responses to change was still obscure made them all the more interesting. By the mid-1830s, of course, most Americans doubtless were familiar with particular harbingers of industrial development like the steamboat, the railroad, or one of several scattered clusters of factories north of the Mason-Dixon line. They also may have known about the new factory systems in Britain or France. But few Americans seem to have grasped the profound long-term significance of these ingenious new artifacts.

Before the Civil War, in fact, the lineaments of the oncoming order were not discernible in the largely under-developed—rural or "wild"—North American landscape. The absence of public awareness was matched by the lack of a name for this distinctive process of change. By 1830, it is true, the first terms designed to serve the purpose—Industrial Revolution and industrialism—had just been coined in France and England, but it would be many years before they gained currency. Yet the lack of nouns for referring to what Karl Polanyi would so aptly call The Great Transformation did not constrain the aesthetic imagination.[4] Quite the contrary; perceptive artists and writers wrapped their relevant thoughts and feelings around the more tangible, visible, audible signs of change: sensory images of steam engines, factories, the telegraph, and locomotives. At some point it became clear to me that this new iconography was a vital conduit to my subject. It enabled me to trace lines of tacit meaning from specific images (or clusters of images) in a literary work to its controlling theme, to the writer's other works, to the work of other writers, and outward to images or events, past and present, that represented the transformation of American society.[5]

Take, for example, Hawthorne's "Ethan Brand," a curiously fragmented allegorical tale, which turns on a variant of the Faustian theme of alienation: the protagonist's debilitating sense of separation from nature, other people, and himself. For the setting, Hawthorne drew extensively upon observations he had made during a walking tour of the Berkshires. Factories with modern machinery recently had been built there, and he was struck by the picturesque impressions they made in the midst of wild scenery. He noted the melancholy women workers staring out of factory windows, and a man whose hand had been cut off by a machine. In composing the story, however, Hawthorne omitted almost all explicit references to the factories, the new artifacts, or their disturbing implications. Yet when we read the published story with his notebook observations in view, it is obvious that the protagonist's alienation was prompted by his disturbed reaction to oncoming industrialism. In the story, indeed, those feelings are bound up with a cluster of less explicit, correlative industrial images: fire, iron, and smoke. By tracing the incidence of that cluster, moreover, I discovered its comparable role in several other works of Hawthorne's, including *The Scarlet Letter*, and—even more revealing—in a thematically related episode of the novel Melville dedicated to him, *Moby-Dick*.[6]

* * * * *

When I submitted my hastily written dissertation in 1949, it was, like most dissertations, far from being a book, and I worked on it intermittently for the next fifteen years. Fifteen years. Even back in 1964, I must admit, that was deemed excessive, but today it is hard to imagine that an untenured scholar would be allowed that much time to get out a first book—at least not if he meant to hold on to an appointment at a respectable college or university. By way of self-exculpation, I might note that I published some essays and did a lot of teaching along the way. But those activities don't count for much these days, when the publish-or-perish dictum routinely is enforced with undiscriminating stringency.

In moving from the thesis to the book, I benefited immeasurably from a continuous, occasionally contentious dialogue—oral and epistolary—with my teacher, sometime colleague, and dear friend Henry Nash Smith.[7] An Americanist fascinated by the efficacy of images and symbols, Smith was a path-breaking student of the reciprocal relations between canonical literature, popular culture, and a putative collective consciousness. In *Virgin Land: The American West as Symbol and Myth* (1950), he ascribed much of American thought and behavior to a shared vision of the nation's future, heritage of biblical myth, as the new Garden of the World. His analysis was persuasive up to a point, but I felt he slighted the challenge to that vision—so conspicuous in the iconography I was assembling—posed by the oncoming industrial order.

I had been struck by the recurrence of a particular episode in the work of many canonical American writers—Emerson, Thoreau, Hawthorne, Melville, Mark Twain, Henry James, Henry Adams, Frank Norris, Ernest Hemingway, William Faulkner—in which a machine, or some other token of the new industrial power, suddenly intrudes upon

the serenity enjoyed by the writer, or a fictional protagonist, situated in a natural, perhaps idyllic, setting. The
event typically arouses feelings of dislocation, anxiety, and
foreboding. Its evocative power derives from the seemingly
universal connotations of the contrast between the industrial machine, say a steam locomotive, and the green landscape. The sensory attributes of the engine—iron, fire,
steam, smoke, noise, speed—evoke the essence of industrial power and wealth, and when set against the attributes
of a natural terrain—fecundity, beauty, serenity, an ineffable numinosity—the effect can be spine-tingling. The
recurrence of the "interrupted idyll" testifies to the
salience of the conflict of meaning and value generated by
the onset of industrial capitalism. It prefigured the emergence of what has proved to be a major cultural divide, separating those Americans who accept material progress as
the primary goal of our society from those who—whatever
their ideals of the fulfilled life—do not.

My first step in turning the dissertation into a book was
to extend the scope of the industrial iconography to the
larger culture. Thus far I had culled my examples almost
exclusively from literary sources, but now, with the help of
a research grant and an assistant, I collected a large sample
of technological images from paintings, newspapers, magazines, folklore, political debates, and ceremonial oratory
between 1830 and 1860. About half of the images—
between three and four hundred—were merely factual,
descriptive, or, in effect, intellectually and emotionally
neutral. I weeded them out and focussed on those which
conveyed a semblance of thought and feeling. To my surprise, almost all of the remaining images also depicted the
new technological artifacts *in a landscape*—a natural setting
either wild or rural; quite a few in fact were variants of the
machine-in-the-landscape trope.

This discovery indicated just how closely the meaning of the new industrial power was bound up, in the minds of Americans, with its perceived relation to nonhuman nature. But it also revealed a striking divergence between the responses characteristic of two spheres of the culture. In the "high" literary sphere, they tended (as in the episode of the "interrupted idyll") to be negative in spirit, but in the wider, more popular sphere they generally were affirmative, even celebratory. In newspapers, popular magazines, and political speeches of the era, the new technological artifacts won praise as emblems of the "conquest of nature," or of America's "Manifest Destiny" to occupy the continent, or—most often and most fulsomely—of the prevailing faith in the idea that history is a record of steady, continuous, cumulative material Progress. Spokesmen for the dominant economic and political elites, primarily journalists, businessmen, and politicians (popular orators like Henry Clay, Edward Everett, and Daniel Webster), fashioned a hyperbolic, somewhat hackneyed "rhetoric of the technological sublime." They redirected an idiom formerly reserved for the contemplation of Nature's awesome power and divinity to the new technology as the appointed instrument for converting an untamed wilderness into a garden-like social order.[8]

My belated recognition of the singular reach and resonance of the machine-in-the-landscape image—its efficacy in representing opposed responses to the emergence of industrial capitalism—also gave me a title. "The Machine in the Garden" was particularly appealing because, for one thing, it summed up my debt to—and my differences with—Henry Nash Smith and *Virgin Land*. When yoked together, moreover, the two "root metaphors" became a third distinct and remarkably inclusive bipolar metaphor of contradiction.[9] ("Contradiction" in that it served equally well to express the typically opposed, even irreconcilable reactions of the dominant and the adversary cultures.) Of course each image had had a rich history in its own right. The chief referent of the "machine," as I first conceived of it, was simply the iron machinery of the industrial revolution. But that literal, workaday sense of machine had had a semantic precursor in the seventeenth-century trope of a metaphysical Machine—emblem of a materialist and mechanistic cosmology—which had been pivotal in the scientific and philosophic revolution associated with Newton, Descartes, and Bacon. The dual meaning of Machine—a cosmos as well as an industrial tool—would be countered, after the mid-eighteenth century, by the organic, biological images of nature favored by exponents of the Romantic recoil from triumphant "mechanism," comprising the new science, technology, and philosophic rationalism. It is not surprising, under the circumstances, that the sudden appearance of the machine technology in the underdeveloped "new world" setting provided the sensory components of a—perhaps the—great central figurative conception of nineteenth-century American culture.

In 1956 I published an essay, "The Machine in the Garden," in which I summarized the argument of my half-finished book.[10] By that time I also was becoming aware of the intriguing affinities between the American works in which the interrupted idyll occurs and Virgilian pastoral. At first, admittedly, it seemed odd, even paradoxical, that our gifted nineteenth-century writers had responded to the invention of the novel artifacts of the Industrial Revolution by reverting to—or instinctively reinventing—the pastoral, one of Western culture's oldest, most enduring modes of thought and expression. They might have been expected, like the modernist artists and writers at the turn of the twentieth century, who had responded to the "Second Industrial Revolution" with an astonishing outburst of formal experimentation, to have devised a correspondingly novel set of expressive forms. As it turned out, however, the adaptation of the pastoral mode was ideally suited to the expressive purposes of American writers.

The pertinence of pastoralism to their work is most apparent in their delineations of the native landscape. This first became evident when I juxtaposed the design of the landscape in passages by Hawthorne and Virgil. The setting Hawthorne creates for a casual notebook account of his own experience of an interrupted idyll almost exactly replicates the tripartite symbolic terrain of Virgil's first Eclogue. The poem's topography is divided into: (1) an off-stage center of power in Rome; (2) an adjacent stretch of wild, untrammeled nature; and between them (3) an idealized "middle landscape," neither wild nor overcivilized, where the dream of harmony between humanity and nature might be attainable. Each sector has its counterpart in the Hawthorne passage and, as it turned out, in many works by many American writers. In the symbolic geography of America embodied in such books as Jefferson's *Notes on Virginia*, Thoreau's *Walden*, Melville's *Moby-Dick*, and Twain's *Huckleberry Finn*, each sector of the Latin poet's characteristic landscape has its Euro-American counterpart. Thus the Eastern seaboard of the United States, or some stereotypical American idea of Europe, replaces Virgil's Rome as the emblematic locus of complexity, wealth and power; the Western frontier stands in for the wild, untamed border of Virgil's pasture; and Virginia (or Walden Pond, or the musky meadow Ishmael conjures on the deck of the Pequod, or Huck and Jim's raft on the Mississippi) takes the place of the Latin poet's Arcadian pasture.

The narratives of the hybrid genre, the American pastoral romance, typically are played out in some variant of this symbolic setting. The action is set in motion by a pastoral impulse—the protagonist's familiar urge, in the face of the established order's growing complexity and power, to get away. "Back out of all this now too much for us," is Robert Frost's unmatched evocation of the impulse.[11] The pastoral figure retreats from this alienating situation to a terrain marked by fewer signs of human intervention. (The retreat also is a quest, a journey in search of an alternative form of life "closer to nature.") In the ideal type of this modern variant of the perilous journey, three defining episodes recur. One is a moment of ecstatic fulfillment, perhaps shared with a companion or lover, when the protagonist enjoys a feeling of transcendent harmony with his surroundings; this epiphany, which often brings in aspects of the Protestant conversion experience, is of course fleeting, and issues in what Frost calls "a momentary stay against confusion."[12] It is striking that so many of the conspicuous voices and characters in American literature tend to join the recovery of self with the recovery of the natural, and to represent their deepest longings in numinous visions of landscape. The second episode is the pastoral figure's thrilling, tonic, yet often terrifying encounter with wildness: some aspect, external or subjective, of unmodified, intractable, or hostile nature. Here, as in the idyll, the narrative in effect arrests the centrifugal impulse with which it began. So does the third episode: the sudden appearance of the machine in the landscape. A distinctive industrial-age feature of the pastoral mode, this intrusive artifact figures forth the unprecedented power and dynamism of the oncoming order, and it exposes the illusory character of the retreat to nature as a way of coping with the ineluctable advance of modernity. As Thoreau

notes on hearing the locomotive's shrill whistle in the
Walden woods, "We have constructed a fate, an *Atropos*,
that never turns aside."[13]

When I began my research, I wanted to find out which
American, if any, deserved credit for having been the first
to imagine the profundity of the change initiated by the
new industrial power. I eventually settled on two candi-
dates: Tench Coxe, Alexander Hamilton's brilliant assistant at the Treasury who wrote much, perhaps most, of the
prophetic 1791 *Report on Manufactures*, and Thomas Jefferson. What particularly recommends Jefferson's Query
XIX, a chapter of his only book, *Notes on Virginia* (1785), is
the distinctively pastoral cast of his argument against introducing British-style manufacturing in the United States.
There he momentarily envisages (he partly retracts it later)
an American society based on three principles, at once pastoral and ecological—principles wholly irreconcilable with
what even then was coming to be the dominant American
ideology of material progress. Anticipating late twentieth-
century environmentalism, Jefferson envisions an America
which would (1) restrict its institutions and practices to
those consonant with the opportunities and constraints
inherent in nonhuman nature; (2) take as its primary economic goal equitable sufficiency (rather than the maxi-
mization of wealth and power); and which, in setting social
goals, would (3) subordinate economic to other less tangible, quality-of-life criteria of public well-being.[14]

From the time of its origin in the ancient Near East, the
hallmark of pastoral had been the contrast, explicit or
tacit, between two ways of life: one complex and the other
simple; one oriented toward power, wealth, and status, the
other toward love, art, and music; and each grounded in a
distinct set of relations with the natural world. The singular
pertinence of the mode to the experience of Europeans in

colonizing remote lands was not lost on Shakespeare, who conceived of *The Tempest* as a version of pastoral. The point of my chapter on that play, then, was to establish the close affinity between that ancient but revitalized form and the myth of America as a new beginning. In their contrasting views, Prospero and Gonzalo prefigured the divergence between the progressive and pastoral responses of Americans to the onset of industrialism.

Like most historical inquiries, *The Machine in the Garden* probably reveals as much about the era in which it was written as the one it was written about. In 1961, three years before it appeared, Rachel Carson published *Silent Spring*, a book said to mark the inception of today's environmental movement. Carson's opening sentence introduces a quintessential American fable: "There was once a town in the heart of America," she begins, "where all life seemed to live in harmony with its surroundings." She lyrically describes the serene seasonal round ("foxes barked in the hills and deer silently crossed the fields, half hidden in the mists of fall mornings"), and then, with a sharp change of mood, the abrupt intrusion of an external force:

> Then a strange blight crept over the area and everything began to change. Some evil spell had settled on the community; mysterious maladies. . . . Everywhere was a shadow of death. . . . There was a strange stillness. The birds, for example,—where had they gone? . . . It was a spring without voices.

In this variant of the machine-in-the-garden trope, the initial surrogate for the machine is "a white granular powder" falling from the sky, a substance Carson associates, a few pages later, with Strontium 90, a by-product of nuclear explosions. In the body of the book, however, she makes DDT and other deadly pesticides, as well as the entire chemical industry, exemplars of modern industrial society's "war against nature." That the founding text of the emerging environmental movement was framed in this way suggests the prominent role played by pastoral ideals in general, and by the machine-in-the-garden trope in particular, in the rhetoric accompanying the imminent explosion of countercultural activism.

At first *The Machine in the Garden* was well received inside and outside the academy. It was translated into several languages, and it was cited approvingly by scholars in fields other than history and literature—notably art history, architecture, landscape architecture, and planning. Within American studies, however, it eventually came under severe attack. In 1972, Bruce Kuklick, a philosopher and an Americanist, condemned it along with several other books— Smith's *Virgin Land* (1950), R. W. B. Lewis's *The American Adam* (1955), John William Ward's *Andrew Jackson: Symbol for an Age* (1955), and Alan Trachtenberg's *Brooklyn Bridge* (1965)—as an example of the inexact scholarship practiced by adherents of what he called the "Myth and Symbol" school.[15] Kuklick charged all of us, whose project he rightly identified as a kind of history, with generalizing about American thought and behavior on the basis of insufficient empirical evidence; with assuming a Cartesian split between thought and material reality; and with imputing excessive causal efficacy to ideas, especially free-floating literary symbols and myths. It was a trenchant, plausible, influential critique. "Myth and Symbol School" caught on as a derogatory watchword, and at least one scholar accused its adherents of mounting a conspiracy.[16] To this day embattled Americanists habitually invoke "myth-and-symbol" as a dismissive epithet of self-evident validity.

In the event, Kuklick's essay proved to be the leading edge of a wave of revisionary criticism that engulfed American studies, along with all the humanistic disciplines, later in the 1970s. Partly due to the political frustration felt by academic intellectuals after the dissipation—some would say defeat—of the 1960s radical Movement, and partly in keeping with the related philosophic "turn" to language, deconstruction, critical theory, and postmodern multiculturalism, books like *The Machine in the Garden* came to be seen by a younger scholarly generation—with some justice—as the expression of a timid, elitist, white male mentality in the service of an entrenched establishment. Since then scholarship in American studies has been dominated by a commitment to the primacy, in the study of American society and culture, of *difference*—of the marked differences, that is, in the experience of Americans as determined by their class, ethnicity, race, gender, or sexual preference. In my judgment, no fair-minded scholar could fail to acknowledge the enriching, liberating, and—all in all—the largely salutary consequences of this intellectual and moral turnabout.

The ideological swerve of the 1970s radically altered the standpoint from which scholars in American studies approach the past. It opened up vast, relatively unexplored areas of knowledge about the history and the expressive life—in literature, art, music—of women, ethnic and racial minorities, gays, and members of the working class. Books written before, say, 1975, inevitably betray the limitations of what now appears to have been a blinkered time. Today, for instance, it would not be possible for me to write certain sentences in *The Machine in the Garden* that tacitly generalize about the thought or behavior of "Americans," unqualified by the explicit distinctions that an informed multicultural consciousness—and conscience—now would compel me to recognize. If I were to write a new version of the book today, I surely would want to include the work of several talented but hitherto insufficiently valued practitioners of the pastoral mode, such as Sarah Orne Jewett, Willa Cather, and Jean Toomer.

To many exponents of the new critical dispensation, *The Machine in the Garden,* like other books of its time, exemplifies an allegedly "holistic," universalizing tendency. They charge us—their authors—with depicting American society and culture as if it constituted a single unified whole. This charge, in my opinion, is grossly overstated and frequently inaccurate. True, the divisions in American society and culture highlighted by the earlier generation of scholars were not those on which today's multiculturalist scholars choose to focus. Nevertheless, *The Machine in the Garden* emphasizes a fundamental divide in American culture and society. It separates the popular affirmation of industrial progress disseminated by spokesmen for the dominant economic and political elites, and the disaffected, often adversarial viewpoint of a minority of political radicals, writers, artists, clergymen, and independent intellectuals. It is a split between those who accept the primacy of material progress, and those who emphasize the less tangible aesthetic, moral, and environmental "qualities of life." Granted that I did not establish this distinction with social scientific precision, its central place in my analysis surely belies the charge that *The Machine in the Garden* advances a simplistic universalizing concept of American society and culture.

The fact is that I myself had underestimated the depth and durability of the conflict I wrote about. Though the chronological center of *The Machine in the Garden* is the mid-nineteenth century, I closed the book in the present tense with the implication that today (1964), in the era of high technology, the pastoral response to industrial society almost certainly had become anachronistic, and that it soon might be expected to lose its hold on the minds of disaffected Americans. That tacit prediction hardly could have been more quickly contradicted by events. On December 2, 1964, a few weeks after the book was published, the Berkeley student rebellion began. It was led by young radicals who had been active in the Civil Rights campaign in the South, and the manifest continuity between the rhetoric of their spokesman, Mario Savio ("You've got to put your bodies upon the [machine] and make it stop") and that of Henry Thoreau in his 1849 essay on "Civil Disobedience" ("Let your life be a counter friction to stop the machine") turned out to be the mere surface expression of a much deeper ideological continuity between nineteenth-century pastoralism and the radical Movement (or counter-culture) of the 1960s.

Indeed, Savio's speech rekindled the charged language of the conflict at the heart of *The Machine in the Garden*. He intended his image of a machine to conjure the military-industrial complex, and he in effect asked his fellow students to mobilize a force of nature—their own frail bodies—in a last desperate gesture aimed at restraining the misuse of organized power. During the next ten years, while the United States waged a high technology war against an Asian peasant society, expressions of hostility to the "machine"—and calls to "Stop the Machine!"—remained battle cries of the radical Movement. It was another strident explosion of the adversary culture which had survived, with occasional outbursts, since its initial appearance in Thoreau's day more than a century earlier.[17] Revulsion at the technocratic direction of American life was a primary spur to the radicals of the 1960s, whose aspirations for their society were summed up by Charles

Reich in the inspired title of his 1970 book, *The Greening of America.*

Since the Vietnam era, this collective mentality has informed many of the dissident movements adhered to by discontented Americans. I have in mind environmentalism, the antinuclear movements (against both nuclear power and nuclear weaponry), the voluntary simplicity movement, as well as "green" tendencies within the feminist, gay rights, Native American, African-American and Hispanic movements. Similar views find expression in the widespread inclination of privileged but morally troubled Americans to "recover the natural," and to repudiate the Calvinist work ethic in favor of a "new" set of rules—an ethic of "self-fulfillment." These political movements and tendencies represent a marked departure from the dissident politics characteristic of the pre–World War II era. They differ from the trade union, populist, progressive, or New Deal movements in that they do not arise from, or centrally involve, antagonistic economic interests. As noted by one prominent student of these "new social movements," the chief conflict they represent is not so much that which sets workers against owners, as that which sets a population against the "apparatus" which appears to dominate it.[18] Could it be that this kind of oppositional politics is what the future now holds in store? If so, it is enlightening to recall that its emergence was foretold by American writers in their initial responses to the appearance of the new machine power in the national landscape.

Notes

Wherever possible I have indicated the approximate location of quoted passages in my text. In referring to standard works for which many editions are available, I have cited the chapter rather than the pages of a particular edition.

THE FOLLOWING ABBREVIATIONS ARE USED IN CITING PERIODICALS:

AHR	American Historical Review
AL	American Literature
Am. J. of Sci.	American Journal of Science
AQ	American Quarterly
Dem. Rev.	United States Democratic Review
JEH	Journal of Economic History
JHI	Journal of the History of Ideas
MLN	Modern Language Notes
MLR	Modern Language Review
MVHR	Mississippi Valley Historical Review
NAR	North American Review
NEQ	New England Quarterly
Sci. Am.	Scientific American
SP	Studies in Philology

CHAPTER I

1. William Empson, *Some Versions of Pastoral*, London, 1950; for my conception of pastoralism I also am indebted to: C. L. Barber, *Shakespeare's Festive Comedy: A Study of Dramatic Form and Its Relation to Social Custom*, Princeton, 1959; Richard J. Cody, "The Pastoral Element in Shakespeare's Early Comedies," unpubl. diss., University of Minnesota, 1961; Walter W. Greg, *Pastoral Poetry and Pastoral Drama*, London, 1906; Frank Kermode, "Introduction," *English Pastoral Poetry: From the Beginnings to Marvell*, London, 1952, and "Introduction," William Shakespeare, *The Tempest*, Arden Edition, 5th ed. revised, London, 1954; Erwin Panofsky, "*Et in Arcadia Ego:* Poussin and the Elegiac Tradition," *Meaning in the Visual Arts: Papers in and on Art History*, New York, 1957; Adam Parry, "Landscape in Greek Poetry," *Yale Classical Studies*, XV (1957), 3–29; Renato Poggi-

oli, "The Oaten Flute," *Harvard Library Bulletin*, XI (Spring, 1957), 147–84; Hallett Smith, "Pastoral Poetry," *Elizabethan Poetry: A Study in Conventions, Meaning, and Expression*, Cambridge, Mass., 1952, pp. 1–63; Bruno Snell, "Arcadia: The Discovery of a Spiritual Landscape," *The Discovery of the Mind: The Greek Origins of European Thought*, trans. T. G. Rosenmeyer, Cambridge, Mass., 1953.

2. Richard Hofstadter, *The Age of Reform: From Bryan to F.D.R.*, New York, 1955; Marvin Meyers, *The Jacksonian Persuasion: Politics and Belief*, New York, 1960; Henry Nash Smith, *Virgin Land: The American West as Symbol and Myth*, Cambridge, Mass., 1950.

3. José Ortega y Gasset, *The Revolt of the Masses*, trans. anon., London, 1932, p. 89.

4. *A General Introduction*, trans. Joan Riviere, New York, 1920, p. 325; *Civilization and Its Discontents*, trans. Joan Riviere, London, 1930, p. 44.

5. *The American Notebooks*, ed. Randall Stewart, New Haven, 1932, pp. 102–5.

6. *The Complete Stories of Herman Melville*, ed. Jay Leyda, New York, 1949, p. 198.

7. *Journals of Ralph Waldo Emerson*, ed. E. W. Emerson and W. E. Forbes, Boston, 1909–14, VI, 322; for the concept of convention used here, see Harry Levin, "Notes on Convention," *Perspectives of Criticism*, Harvard Studies in Comparative Literature, Cambridge, 1950.

8. "On the Projected Kendal and Windermere Railway," *The Poetical Works of Wordsworth*, ed. Thomas Hutchinson, London, 1904, pp. 282–3.

9. *The Poetical Works of William Blake*, ed. John Sampson, London, 1913, p. 370; Alfred North Whitehead, *Science and the Modern World*, New York, 1947, p. 138.

10. Snell, "Arcadia"; *Virgil, The Pastoral Poems*, trans. E. V. Rieu, Penguin Books, Harmondsworth, England, 1949. Quoted by permission of the publisher.

11. Arthur O. Lovejoy, et al., *A Documentary History of Primitivism and Related Ideas*, Baltimore, 1935, p. 369; Lovejoy saw the roots of cultural primitivism as "various and incongruous." "Common to them all, indeed, is the conviction that the time — whatever time may, for a given writer, be in question — is out of joint; that what is wrong with it is due to an abnormal complexity and sophistication in the life of civilized man, to the pathological multiplicity and emulativeness of his desires and the oppressive over-abundance of his belongings, and to the factitiousness and want of inner spontaneity of his emotions; that 'art,' the work of man, has corrupted 'nature,' that is, man's own nature; and that the model of the normal individual life and the normal social order, or at least a nearer approximation to it,

is to be found among contemporary 'savage' peoples, whether or not it be supposed to have been realized also in the life of primeval man. Civilized man has been almost continuously subject to moods of revolt against civilization, which in some sense is, indeed, profoundly contrary to his nature; and in the serious preachers of primitivism this revolt has been chronic and intense. But the belief in the superiority of the simple life of 'nature' has been the manifestation sometimes of a hedonistic, sometimes of a rigoristic and even ascetic, conception of the nature of the good, and sometimes a mixture of both." Foreword to Lois Whitney, *Primitivism and the Idea of Progress in English Popular Literature of the Eighteenth Century*, Baltimore, 1934, pp. xiv–xv.

12. Panofsky, *"Et in Arcadia Ego."*

13. W. W. Rostow, *The Stages of Economic Growth, A Non-Communist Manifesto*, Cambridge, 1960, p. 7, and chart, p. xii.

14. Ralph Waldo Emerson, "Art," *Essays, First Series*, in *Complete Works*, 11 vols., Boston, 1885, II, 328.

15. Robert Frost, "The Figure a Poem Makes," *Complete Poems*, New York, 1949, p. vi.

CHAPTER II

1. For a general survey, see Robert Ralston Cawley, *The Voyagers and Elizabethan Drama*, Boston, 1938; and the same author's "Shakespeare's Use of the Voyagers in *The Tempest*," *PMLA*, XLI (1926), 688–726; more recently Frank Kermode has reviewed the evidence in the revised Arden Edition of *The Tempest*, London, 1954, pp. xxv–xxxiv; Ariel: I, ii, 229. I am following the Arden text.

2. "The first voyage made to the coasts of America . . ." in Richard Hakluyt, *The Principle Navigations Voyages Traffiques and Discoveries of the English Nation*, Glasgow, 1904, 12 vols., VIII, 297–310.

3. Kenneth Clark, *Landscape into Art,* Edinburgh, 1956, Ch. I.

4. My discussion of utopianism draws upon George Kateb, *Utopia and Its Enemies*, New York, 1963; David Potter, *People of Plenty: Economic Abundance and the American Character*, Chicago, 1954.

5. Strachey, "A true reportory of the wracke, and redemption of Sir Thomas Gates Knight; upon, and from the Ilands of the Bermudas . . . ," *Hakluytus Posthumus or Purchas His Pilgrimes*, ed. Samuel Purchas, Glasgow, 1906, 20 vols., XIX, 6, 12; Cawley, *The Voyagers*, p. 347ff, and Richard Bernheimer, *Wild Men in the Middle Ages: A Study in Art, Sentiment, and Demonology*, Cambridge, Mass., 1952.

6. *History of Plymouth Plantation, 1606–1646*, ed. William T. Davis, New York, 1908, pp. 94–6.

7. The concept of "root metaphor" is developed by Stephen C. Pepper in *World Hypotheses, A Study in Evidence*, Berkeley, 1942; Emerson, *Complete Works*, 11 vols., Boston, 1885, V, 52.

8. For the way American conditions nurtured the utilitarian germ at the center of Puritan thought, see Perry Miller, *The New England Mind: The Seventeenth Century*, New York, 1939, especially pp. 393ff; Miller's views are extended by Alan Heimert, "Puritanism, the Wilderness, and the Frontier," *NEQ*, XXVI (1953), 361–82; Robert K. Merton has demonstrated the connections between English Puritanism and the development of technologically advanced capitalism in his "Science, Technology, and Society in Seventeenth-Century England," *Osiris*, Bruges, 1938, IV, II.

9. *A Discovery of the Barmudas*, ed. Joseph Quincy Adams, Scholars' Facsimiles and Reprints, New York, 1940, pp. 8ff, first published in 1610.

10. "A true reportory," Purchas, XIX, 13; anon., "A True Declaration of the estate of the Colonie in Virginia, with a confutation of such scandalous reports as have tended to the disgrace of so worthy an enterprise . . . , London, 1610," *Tracts and Other Papers*, ed. Peter Force, Washington, 1844, 4 vols., III, 14.

11. II, i, 34–51.

12. *The Voyages of Christopher Columbus*, The United States Catholic Historical Society, New York, 1892, p. 48.

13. II, i, 143–60.

14. Montaigne's "Of Cannibals" is the only undisputed source for the play; Kermode discusses the relationship in his edition, pp. xxxivff; T. P. Harrison, Jr., compares the two texts in detail, noting that Shakespeare, "by a series of touches . . . accentuates the hollowness of this ideal," "Aspects of Primitivism in Shakespeare and Spenser," *Studies in English*, University of Texas Publication 4026 (1940), 37–71; A. O. Lovejoy refers to Gonzalo's vision as expressive of "Shakespeare's own extreme antipathy to the [Montaigne] passage," "On the Discrimination of Romanticisms," *Essays in the History of Ideas*, Baltimore, 1948, p. 238n; that the play embodies a critique of sentimental primitivism is beyond argument, but to call Shakespeare's attitude "extreme antipathy" is an exaggeration; Gonzalo emerges a sympathetic figure, and all the evidence indicates that Shakespeare is reaffirming an ideal resolution of the tension between art and nature; II, i, 163–4.

15. I, ii, 149–51.

16. I, i.

17. I, ii, 269; I, ii, 333–4; I, ii, 281–4.

18. V, i, 64.
19. II, ii, 38.
20. III, iii, 27–34; I, ii, 347–50; I, ii, 353–5.
21. For the technological overtones of Prospero's magic, see G. L. Kittredge, "Introduction," *The Tempest*, Ginn and Co., Boston, 1939, p. xviii; Prospero, says Kittredge, uses art as "a method of controlling the forces of nature." To the Elizabethan audience he "was as comprehensible in his feats of magic as a chemist or an electrical engineer is to us moderns."
22. III, ii, 89–92; V, i, 25–7; John Berryman called attention to Prospero as the embodiment of Hamlet's ideal in a lecture, several years ago, at the University of Minnesota.
23. Critics who deny the presence of a positive view of nature in the play include Elmer Edgar Stoll, "The Tempest," *PMLA*, XLVII (1932), 699–726, and T. P. Harrison, Jr., "Aspects of Primitivism"; I, ii, 390–96; II, i, 45, 51.
24. Knight, *The Shakespearian Tempest*, 3rd ed., London, 1953, p. 17; III, ii, 133–6; II, ii, 160–72.
25. V, i, 18–27.
26. IV, i, 60–63; IV, i, 110–17; II, i, 158–60; V, i, 62; 68–9.
27. V, i, 181–4.
28. The term "meliorist" will serve to distinguish Prospero's attitude from that of the nineteenth-century "believer" in progress — that is, one who believed in history as the record of inevitable improvement in the condition of mankind. For the Baconian aspect, see J. B. Bury, *The Idea of Progress: An Inquiry into Its Origin and Growth*, London, 1921, pp. 50–63.
29. IV, i, 122–4; the stage direction follows IV, i, 138; IV, i, 139–40.
30. IV, i, 188–9; though we have no adequate study of Shakespearian pastoralism, some useful ideas may be found in: Edwin Greenlaw, "Shakespeare's Pastorals," *SP*, XIII (1916), 122–54; Thomas P. Harrison, Jr., "Aspects of Primitivism"; Edgar C. Knowlton, "Nature and Shakespeare," *PMLA*, LI (1936), 717–44; Richard Cody has begun what promises to be the most illuminating study (see Ch. I, note 1 above).
31. *The Winter's Tale, Shakespeare's Twenty-Three Plays and the Sonnets*, ed. Thomas Marc Parrott, New York, 1953, IV, iv, 83–97.
32. V, i, 274–6; Emerson, *Journals*, III (1833), 207.
33. V, i, 208–13.
34. Boris Pasternak, "Translating Shakespeare," trans. Manya Harari, *I Remember, Sketch for an Autobiography*, New York, 1959, p. 148.

CHAPTER III

1. Louis B. Wright, "Robert Beverley II: Historian and Iconoclast," *The First Gentlemen of Virginia*, San Marino, 1940, pp. 286–311; see also Wright's introduction to Robert Beverley, *The History and Present State of Virginia*, ed. Louis B. Wright, Chapel Hill: University of North Carolina Press, 1947, pp. xi–xxxv; pp. 15–16.

2. Beverley, p. 17.

3. Beverley, p. 297; p. [115]; p. 140.

4. See, e.g., Maurice A. Mook, "The Anthropological Position of the Indian Tribes of Tidewater, Virginia," *William and Mary Quarterly*, 2nd series, XXIII (1943), 27–40, and John R. Swanton, *The Indians of the Southeastern United States*, Smithsonian Institution Bureau of American Ethnology Bulletin 137, U. S. Printing Office, 1946. Roy Harvey Pearce discusses Beverley's attitude toward the Indians in *The Savages of America: A Study of the Indian and the Idea of Civilization*, Baltimore, 1953, pp. 42–3; Beverley, p. 51; p. 38; p. 171.

5. Beverley, p. 156.

6. Beverley, p. 156; p. 233; p. 296.

7. Beverley, p. 298; Myra Reynolds probably established the idea of Thomson as the "first adequate exponent" of the new attitude toward nature in her pioneering study, *The Treatment of Nature in English Poetry, Between Pope and Wordsworth*, Chicago, 1909; see esp. pp. 58ff; for the idea that American writers lagged behind the British in this regard, see Mary E. Wooley, "The Development of the Love of Romantic Scenery in America," *AHR*, III (1897), 56–67; Beverley, p. 298.

8. Beverley, p. 298–9; p. 316.

9. Mircea Eliade, "The Yearning for Paradise in Primitive Tradition," *Daedalus* (Spring, 1959), pp. 255–67, and *Cosmos and History: The Myth of the Eternal Return*, trans. Willard R. Trask, New York, 1959.

10. Anon., *A Letter from South Carolina . . .* (dated June 1, 1710) *Written by a Swiss Gentleman to His Friend at Bern* (London, 1732, 2nd ed.), quoted by Michael Kraus, "America and the Utopian Ideal in the Eighteenth Century," *MVHR*, XXII (1936), p. 490.

11. For the idea of the single art of landscape, see Christopher Hussey, *The Picturesque*, London, 1927; Kenneth Clark, *Landscape into Art*, Edinburgh, 1956, pp. 77–8.

12. Alexander Pope, *Pastoral Poetry and An Essay on Criticism*, eds. E. Audra and Aubrey Williams, London and New Haven, 1961, pp. 59–61; Spence, *Observations, Anecdotes, and Characters of Books and Men*, London, 1820, p. 2.

13. J. E. Congleton summarizes the ideas in *Theories of Pastoral Poetry*

in England, 1684–1798, Gainesville, Florida, 1952; Philips, "The Sixth Pastoral" (1709), lines 37–44, *English Poetry of the Eighteenth Century,* ed. Cecil A. Moore, New York, 1935, p. 86.

14. *The Works of Joseph Addison,* New York, 1855, VI, 324.

15. *Works,* VI, 337–8.

16. *Works,* VI, 338–9.

17. For the parallels between "The Seasons" and Virgil's *Eclogues* see Elizabeth Nitchie, *Vergil and the English Poets,* New York, 1919, pp. 179–96; also, Congleton, *Theories of Pastoral,* pp. 131–2; Thomson, "Spring," lines 95–107, *English Poetry of the Eighteenth Century,* ed. Moore, p. 376; Joseph Warton, *Essay on Pope,* quoted in Robert Kilburn Root, *The Poetical Career of Alexander Pope,* Princeton, 1938, p. 58; *The Literary Bible of Thomas Jefferson, His Commonplace Book of Philosophers and Poets,* ed. Gilbert Chinard, Baltimore, 1928, p. 174.

18. On the influence of science upon attitudes toward landscape see Walter J. Hipple, *The Beautiful, the Sublime, and the Picturesque,* Carbondale, Illinois, 1957; Cecil A. Moore, "The Return to Nature in English Poetry," *Backgrounds of English Literature, 1700–1760,* Minneapolis, 1953; Samuel H. Monk, *The Sublime, A Study of Critical Theories in XVIII-Century England,* Ann Arbor, 1960; Marjorie Hope Nicolson, *Mountain Gloom and Mountain Glory,* Ithaca, 1959; Henry V. S. and Margaret S. Ogden, *English Taste in Landscape in the Seventeenth Century,* Ann Arbor, 1955; Ernest Lee Tuveson, *The Imagination as a Means of Grace,* Berkeley, 1960; Samuel Johnson, *Lives of the English Poets,* ed. George Birkbeck Hill, Oxford, 1905, I, 163–4; Emerson, "Nature," *Complete Works,* 11 vols., Boston, 1885, I, 68.

19. Paul H. Johnstone concisely summarizes the history of the agrarian celebration in "In Praise of Husbandry," *Agricultural History,* XI (1937), 80–95; Hallett Smith takes Sidney's couplet (from "Disprayse of a Courtly Life," 1602) as a paradigm of the "pastoral ethic" in *Elizabethan Poetry: A Study in Conventions, Meaning, and Expression,* Cambridge, Mass., 1952, pp. 10–11.

20. William Empson, *Some Versions of Pastoral,* London, 1950, p. 11.

21. For the "ethic of the middle link" see A. O. Lovejoy, *The Great Chain of Being, A Study of the History of an Idea,* Cambridge, Mass., 1942, esp. pp. 189–207.

22. Hoxie Neale Fairchild, *The Noble Savage, A Study in Romantic Naturalism,* New York, 1928, and A. O. Lovejoy, "The Temporalizing of the Chain of Being," *The Great Chain,* pp. 242–87.

23. In "The Supposed Primitivism of Rousseau's *Discourse on Inequality,*" *Essays in the History of Ideas,* New York, 1960, p. 31.

24. *Lectures on Rhetoric and Belles Lettres,* London, 1785, III, 117–18;

for Blair's influence in America see William Charvat, *The Origins of American Critical Thought, 1810–1835*, Philadelphia, 1936; Jefferson's letter, Feb. 20, 1784, in response to Madison's request of Feb. 11, *The Papers of Thomas Jefferson*, ed. Julian P. Boyd, et al., Princeton, 1952, VI, 537, 544; hereafter, *Papers*.

25. *Works of George Herbert*, ed. F. E. Hutchinson, Oxford, 1941, p. 196; *The Poems and Letters of Andrew Marvell*, ed. H. M. Margoliouth, Oxford, 1952, I, 17; Price, *Observations on the Importance of the American Revolution*, London, 1785, pp. 57–8; Jefferson to Price, Feb. 1, 1785, *Papers*, VII, 630.

26. I am indebted to Professor Robert Elias of Cornell University for calling my attention to *The Golden Age*.

27. "Letter I," *Letters from an American Farmer*, ed. Warren Barton Blake, Everyman's Library, London, 1912.

28. Blake, "Introduction" to Crèvecœur's *Letters*, p. xx.

29. "Letter I," "Letter II."

30. In "Letter I" the minister defines the Farmer's mentality in Lockian terms: "your mind is what we called at Yale College," he says, "a *Tabula rasa*, where spontaneous and strong impressions are delineated with facility"; Lawrence, *Studies in Classic American Literature*, New York, 1951, ch. III; "Letter II."

31. "Letter III"; Lawrence, *Studies*, ch. III; "Letter II."

32. "No great manufacturers" in "Letter III"; "Letter II"; "Letter VIII."

33. The Natural Bridge is described in "Query V," *Notes on the State of Virginia*, ed. William Peden, Chapel Hill: University of North Carolina Press, 1955, p. 24; Locke's *Essay Concerning the True Original Extent and End of Civil Government*, in *The English Philosophers from Bacon to Mill*, ed. Edwin A. Burtt, New York, 1939, p. 422.

34. "Query XI," p. 93.

35. Jefferson to Adams, Jan. 21, 1812, *The Adams-Jefferson Letters, The Complete Correspondence Between Thomas Jefferson and Abigail and John Adams*, ed. Lester J. Cappon, 2 vols., Chapel Hill, 1959, II, 291; Adams to Jefferson, Feb. 3, 1812, II, 296 (". . . I would rather be the poorest Man in France or England . . . than the proudest King . . . of any Tribe of Savages in America"); "Query VI."

36. Jefferson to Adams, June 27, 1813, *Adams-Jefferson Letters*, II, 335.

37. Renato Poggioli, "The Oaten Flute," *Harvard Library Bulletin*, XI (Spring, 1957), 150–53; see also A. Whitney Griswold's analysis of "The Jeffersonian Ideal" in *Farming and Democracy*, New Haven, 1952, pp. 18–46; Griswold does not invoke the concept of pastoralism, but he argues (p. 30) that to Jefferson agriculture "had a sociological rather than an economic value."

38. Empson, *Some Versions of Pastoral*, p. 11; p. 15.

39. Letter to Peter Carr, Aug. 10, 1787, *Papers*, XII, 15.

40. Letter to John Bannister, Jr., Oct. 15, 1785, *Papers*, VIII, 635–7.

41. Constance Rourke, *American Humor, A Study of the National Character*, New York, 1931, p. 15; Lawrence to Amy Lowell, Aug. 23, 1916, in S. Foster Damon, *Amy Lowell, A Chronicle*, Boston, 1935, pp. 370–71.

42. Letter from van Hogendorp, Sept. 8, 1785, *Papers*, VIII, 501–5; Jefferson's reply, Oct. 13, 1785, VIII, 631–4.

43. Hofstadter, *The American Political Tradition and the Men Who Made It*, New York, 1948, pp. 24–5.

44. Van Zandt, *The Metaphysical Foundations of American History*, The Hague, 1959, chs. VIII, IX.

45. Letter to John Randolph, Aug. 25, 1775, *Papers*, I, 240–42; letter, addressee unknown, Mar. 18, 1793. *The Writings of Thomas Jefferson*, ed., Andrew A. Lipscomb, Washington, D.C., 1903, IX, 44–6; letter to Monsieur D'Ivernois, Feb. 6, 1795, *Writings*, IX, 297–301. For a more complete survey of letters on this theme, see Van Zandt, *Metaphysical Foundations*, pp. 189–94.

46. Letter to Charles Wilson Peale, Aug. 20, 1811, *Writings*, XIII, 78–80.

47. Letter to Benjamin Austin, Jan. 9, 1816, *Writings*, XIV, 387–93.

48. Griswold discusses Jefferson's "incomplete" and "reluctant" conversion to manufactures in *Farming and Democracy*, p. 35; see also Van Zandt, *Metaphysical Foundations*, pp. 205–37, and Joseph Dorfman, *The Economic Mind in American Civilization*, 1606–1865, New York, 1946, I, 433–46.

49. Merrill D. Peterson, *The Jefferson Image in the American Mind*, New York, 1960, pp. 443–6.

50. Henry Nash Smith, *Virgin Land: The American West as Symbol and Myth*, Cambridge, Mass., 1950, p. 123.

51. *The Selected Letters of John Keats*, ed. Lionel Trilling, New York, 1951, p. 28; Jefferson to William Short, Nov. 28, 1814, *Writings*, XIV, 214.

CHAPTER IV

1. James Boswell, *Life of Johnson*, Oxford Standard Authors Edition, London, 1953, p. 704.

2. Boulton's letter to Watt, H. W. Dickinson, *Matthew Boulton*, Cambridge, 1937, p. 113, quoted in the *Papers of Thomas Jefferson*, ed. Julian P. Boyd, et al., Princeton, 1952, IX, 401n; Jefferson to Thomson, Apr. 22, 1786, *Papers*, IX, 400–401.

3. Smith, *An Inquiry into the Nature and Causes of the Wealth of*

Nations, ed. Edwin Cannan, Modern Library Edition, New York, 1937, pp. 346–8; Adams letter to Franklin, Aug. 17, 1780, *The Works of John Adams,* Boston, 1852, VII, 247.

4. On the state of American manufactures in 1790, see the survey made by Alexander Hamilton and Tench Coxe in preparation for the "Report on the Subject of Manufactures," *Industrial and Commercial Correspondence of Alexander Hamilton,* ed. Arthur Harrison Cole, Chicago, 1928; Samuel Rezneck, "The Rise and Early Development of the Industrial Consciousness in the United States, 1760–1830," *Journal of Economic and Business History,* IV (Aug., 1932), 784–812; Arthur M. Schlesinger, *The Colonial Merchants and the American Revolution, 1763–1776,* New York, 1918; George S. White, *Memoir of Samuel Slater,* Philadelphia, 1836.

5. Jacob Bigelow, *Elements of Technology, Taken Chiefly from a Course of Lectures . . . on the Application of the Sciences to the Useful Arts,* Boston, 1829; on Bigelow's coinage, see Dirk J. Struik, *Yankee Science in the Making,* Boston, 1948, pp. 169–70; for the word "industrialism," see above, pp. 174ff.

6. The only detailed study of Coxe's thought is Harold Hutcheson's *Tench Coxe, A Study in American Economic Development,* Johns Hopkins University Studies in Historical and Political Science, No. 26, Baltimore, 1938; *Memoirs of John Quincy Adams,* ed. Charles F. Adams, IV, 370; Joseph Dorfman, *The Economic Mind in American Civilization, 1606–1865,* New York, 1946, I, 254.

7. For Coxe's relation to the Federal Constitution, see Hutcheson, *Tench Coxe,* pp. 49–76, and the "introductory remarks" to Coxe's various papers in his own collection, *A View of the United States, in a Series of Papers Written at Various Times Between the Years 1787 and 1794,* Philadelphia, 1794.

8. *A View,* pp. 38–41; unless otherwise noted all quotations from Coxe are taken from *A View,* pp. 1–56.

9. Hugo Arthur Meier, "The Technological Concept in American Social History, 1750–1860," unpubl. diss., U. of Wisconsin, 1950; Jeannette Mirsky and Allan Nevins, *The World of Eli Whitney,* New York, 1952; Charles Sanford, *The Quest for Paradise: Europe and the American Moral Imagination,* Urbana, 1961, pp. 155–75.

10. "Letters of Phineas Bond, British Consul at Philadelphia, To the Foreign Office of Great Britain, 1787, 1788, 1789," *Annual Report of the American Historical Association for the Year 1896,* Washington, D.C., 1897, I, 513–656; William R. Bagnall, *The Textile Industries of the United States,* Cambridge, Mass., 1893, I, 70; White, *Memoir of Samuel Slater.*

11. *The Journal of William Maclay, 1789–1791,* New York, 1927, p. 252.

12. Jefferson to Lithson, Jan. 4, 1805, *Writings of Thomas Jefferson,*

Washington, D.C., 1903, XI, 55–6; Michael Chevalier, *Society, Manners, and Politics in the United States: Letters on North America*, ed. John William Ward, New York, 1961, p. 142.

13. For the decisive place of the mechanical-organic distinction in romanticism, see M. H. Abrams, *The Mirror and the Lamp: Romantic Theory and the Critical Tradition*, New York, 1953, esp. pp. 156ff; Howard M. Jones, "The Influence of European Ideas in Nineteenth-Century America," *AL*, VII (1935), 241–73; Morse Peckham, "Toward a Theory of Romanticism," *PMLA*, LXVI (1951), 5–23; Thomson, "A Poem Sacred to the Memory of Sir Isaac Newton"; Berkeley, *Principles*, in *The English Philosophers from Bacon to Mill*, ed. Edwin A. Burtt, New York, 1939, p. 577.

14. "Art and Artists," reprinted from the Brooklyn *Daily Advertiser*, April 3, 1851, in *The Uncollected Poetry and Prose of Walt Whitman*, ed. Emory Holloway, New York, 1921, I, 241–7.

15. Thomas Boyd, *Poor John Fitch, Inventor of the Steamboat*, New York, 1935, pp. 178–9, and Merrill Jensen, *The New Nation, A History of the United States During the Confederation, 1781–1789*, New York, 1950, pp. 152–3.

16. Morris is discussing his own resolution to restrict the suffrage to freeholders; *The Records of the Federal Convention of 1787*, ed. Max Farrand, New Haven, 1911, II, 201ff.

17. Bagnall, *The Textile Industries*, I, 110; Everett, "American Manufactures," address delivered before the American Institute of the City of New York, October 14, 1831, *Orations and Speeches on Various Occasions*, Boston, 1850, II, 94; in this address Everett also describes the parades in celebration of manufactures in 1788, p. 88.

18. Coxe's part in preparing the report may have been greater than is generally recognized; Julian P. Boyd, the only scholar who has been granted access to the Coxe manuscripts, reports the existence of a draft in Coxe's handwriting (Hutcheson, *Tench Coxe*, p. viii); "Report on the Subject of Manufactures," *Industrial and Commercial Correspondence*, ed. Cole, p. 247; the change in attitude is reflected in the responses to Hamilton's inquiry, e.g., Elisha Colt, reporting on the conditions in the Connecticut woolen industry: "The present use of Machines in England gave their Manufacturers immense advantages over us — This we expect soon to remedy"; *Industrial and Commercial Correspondence*, ed. Cole, p. 10.

19. *Industrial and Commercial Correspondence*, ed. Cole, p. 258.

20. "Letter VI," *Essays Aesthetical and Philosophical*, London, 1910, pp. 37–44.

21. *Critical and Miscellaneous Essays*, New York (Belford, Clarke & Co.), n.d., III, 5–30.

22. For the development of the Marxian concept of alienation from

Hegel's theory (*Entfremdung*), see Herbert Marcuse, *Reason and Revolution: Hegel and the Rise of Social Theory*, New York, pp. 34f and pp. 273f; Fromm, *Marx's Concept of Man*, with a translation from Marx's *Economic and Philosophical Manuscripts* by T. B. Bottomore, New York, 1961, p. 44.

23. Marx, "Alienated Labor," in Fromm, *Marx's Concept of Man*, p. 95; Emerson, "Ode, Inscribed to W. H. Channing," *Complete Works*, 11 vols., Boston, 1885, IX, 73; Emerson, "Historic Notes of Life and Letters in New England," *Works*, X, 310–11.

24. *Eros and Civilization, A Philosophical Inquiry into Freud*, Boston, 1955; *Sartor Resartus, The Life and Opinions of Herr Teufelsdröckh*, ed. Charles Frederick Harrold, New York, 1937, p. 164.

25. There is no adequate survey of the impact of industrialization upon American thought in the period before the Civil War; see Arthur A. Ekirch, Jr., *The Idea of Progress in America, 1815–1860*, New York, 1944 (although Ekirch's evidence suggests a close affinity between the idea of progress and technological development, he does not define the relationship); Marvin M. Fisher, "From Wilderness to Workshop: The Response of Foreign Observers to American Industrialization, 1830–1860," unpubl. diss., University of Minnesota, 1958, "The Iconology of Industrialism, 1830–1860," *AQ*, XIII (Fall, 1961), 347–64; Hugo A. Meier, "American Technology and the Nineteenth-Century World," *AQ*, X (1958), 116–30; "Technology and Democracy, 1800–1860," *MVHR*, XLIII (1957), 618–40; Charles L. Sanford, "The Intellectual Origins and New-Worldliness of American Industry," *JEH*, XVIII (1958), 1–16; John E. Sawyer, "The Social Basis of the American System of Manufacturing," *JEH*, XIV (1954), 361–79; George Rogers Taylor, *The Transportation Revolution, 1815–1860*, New York, 1951; Rush Welter, "The Idea of Progress in America: An Essay in Ideas and Method," *JHI*, XVI (1955), 401–15.

26. "Defence of Mechanical Philosophy," *NAR*, XXXIII (July, 1831), 122–36.

27. Anna Bezanson, "The Early Use of the Term 'Industrial Revolution,'" *Quarterly Journal of Economics*, XXXVI (1922), 343–9; George Norman Clark, *The Idea of the Industrial Revolution*, Glasgow, 1953; Herbert Heaton, "Clio's New Overalls," *Canadian Journal of Economics and Political Science*, XX (1954), 467–77.

28. I believe that the phrase "the industrial revolution incarnate" is J. H. Clapham's, but I cannot now find it in his *An Economic History of Modern Britain*, Cambridge, 1926–38; Emerson, *Works*, II, 33.

29. "M. de Tocqueville on Democracy in America," *Edinburgh Review*, October, 1840, in *Dissertations and Discussions: Political, Philosophical, and Historical*, Boston, 1865, II, 148.

30. "Our Times," *Dem. Rev.*, XVI (March, 1845), 235–42; Alexis de Tocqueville, *Democracy in America*, ed. Phillips Bradley, New York, 1946, II, 74; Pope, "Martin Scriblerus, or, Of the Art of Sinking in Poetry," ch. XI, published with the second volume of Pope's prose works, but possibly written by John Arbuthnot; "Poor Inventors," *Sci. Am.*, II (Aug., 1847), 389.

31. Charles Caldwell, "Thoughts on the Moral and Other Indirect Influences of Rail-Roads," *New England Magazine*, II (April, 1832), 288–300.

32. "Excitement of Curiosity," *Sci. Am.*, II (Oct., 1846), 22; Henry T. Tuckerman, "The Philosophy of Travel," *Dem. Rev.*, XIV (May, 1844), 527–39; Hunt's *Merchants' Magazine*, III (Oct., 1840), 287.

33. *Am. J. of Sci.*, XXXVIII (1840), 276–97.

34. "The Utility and Pleasures of Science," *Sci. Am.*, II (Aug., 1847), 381; Ripley, "Rail Road to the Pacific," *Harbinger*, IV (Dec., 1846), 30–31.

35. "The Telescope," *Sci. Am.*, V (Feb., 1850), 184; "Improved Hay-Maker," *Sci. Am.*, II (March, 1860, new series), 216.

36. Charles Fraser, "The Moral Influence of Steam," Hunt's *Merchants' Magazine*, XIV (June, 1846), 499–515.

37. "The Commencement at Yale College," *New England Mag.*, I (Nov., 1831), 407–14; "The Poetry of Discovery," *Sci. Am.*, V (Nov., 1849), 77; J. J. Greenough, "The New York Industrial Palace," *Am. Polytechnic Journal*, II (1853), 157–60.

38. "American Machinery — Matteawan," *Sci. Am.*, V (April, 1850), 253; Raymond Williams, *Culture and Society, 1780–1950*, New York, 1960; "Thoughts on Labor," *The Dial*, I (April, 1841), 497–519.

39. "Statistics and Speculations Concerning the Pacific Railroad," *Putnam's*, II (Sept., 1853), 270–77; J. Blunt, "The Coal Business of the United States," Hunt's *Merchants' Magazine*, IV (Jan., 1841), 62–72.

40. "Knowledge, Inventors and Inventions," *Sci. Am.*, V (May, 1850), 285; Ewbank quoted in *Sci. Am.*, V (Feb., 1850), 173.

41. Reprinted from the *Times, Littell's Living Age*, XXVII (Nov., 1850), 333–5; Marsh, *Address Delivered before the Agricultural Society of Rutland County*, Rutland, Vt., 1848.

42. "The Mechanical Genius of the Ancients Wrongly Directed," *Sci. Am.*, II (June, 1847), 301.

43. Hunt's *Merchants' Magazine*, III (Oct., 1840), 295; "American Genius and Enterprise," *Sci. Am.*, II (Sept., 1847), 397.

44. Lanman, Hunt's *Merchants' Magazine*, III (Oct., 1840), 273–95; this article (and other covert reactions to technological change) discussed by Bernard Bowron, Leo Marx, and Arnold Rose in "Literature and Covert Culture," *AQ*, IX (Winter, 1957), 377–86, reprinted in some-

what revised form in *Studies in American Culture,* eds. Joseph J. Kwiat and Mary C. Turpie, Minneapolis, 1960, pp. 84–95.

45. Perry Miller, "The Responsibility of Mind in a Civilization of Machines," *American Scholar,* XXXI (Winter, 1961), 1–19; Chevalier, Poussin, Bremer, quoted by Marvin Fisher, "From Wilderness to Workshop," pp. 86–7; (see note 25).

46. "Opening of the Northern Railroad," remarks made at Grafton and Lebanon, N. H., *The Writings and Speeches of Daniel Webster,* Boston, 1903, IV, 105–17; Clarence Mondale, "Daniel Webster and Technology," *AQ,* XIV (Spring, 1962), 37–47.

47. Taylor, *Transportation Revolution,* p. 79; Orvis, "Trip to Vermont," *Harbinger,* V (July, 1847), 50–52.

48. Nathaniel Hawthorne, *The Blithedale Romance,* 1852, ch. III.

49. Merle Curti, "The Great Mr. Locke: America's Philosopher, 1783–1861," *Hunt. Lib. Bulletin,* XI (April, 1937), 107–51; Louis Hartz, *The Liberal Tradition in America, An Interpretation of American Political Thought Since the Revolution,* New York, 1955.

50. John W. Ward, *Andrew Jackson: Symbol for an Age,* New York, 1955; Marvin Meyers, *The Jacksonian Persuasion: Politics and Belief,* New York, 1960.

51. Börn, *American Landscape Painting,* New Haven, 1948; Virgil Barker, *American Painting: History and Interpretation,* New York, 1950; Elizabeth McCausland, *George Inness, An American Landscape Painter, 1825–1894,* New York, 1946.

52. Read, *The New Pastoral,* Philadelphia, 1855, p. vi.

53. "American Manufactures," Hunt's *Merchants' Magazine,* V (1841), 141; "The Memphis Convention," *DeBow's Review* (March, 1850), 217–32.

54. "What Is the Golden Age," *Sci. Am.,* V (Dec., 1849), 109.

CHAPTER V

1. Philip Young, "Fallen from Time: The Mythic Rip Van Winkle," *The Kenyon Review,* XXII (Autumn, 1960), 551; Frederick I. Carpenter surveys variants, " 'The American Myth': Paradise (To Be) Regained," *PMLA,* LXXIV (Dec., 1959), 599–606; R. W. B. Lewis, *The American Adam: Innocence, Tragedy, and Tradition in the Nineteenth Century,* Chicago, 1955; F. O. Matthiessen, "The Need for Mythology," *American Renaissance: Art and Expression in the Age of Emerson and Whitman,* New York, 1941, pp. 626–31.

2. Emerson, *Complete Works,* 11 vols., Boston, 1885, I, 19, 45.

3. *Journals of Ralph Waldo Emerson*, ed. E. W. Emerson and W. E. Forbes, Boston, 1909–14, IV (1837), 207; VI (1843), 397.

4. *Journals*, IV (1837), 288; *Works*, I, 14–17; Edwards, "Personal Narrative," in *Jonathan Edwards*, eds. Clarence H. Faust and Thomas H. Johnson, American Writers Series, New York, 1935, pp. 60, 63; Perry Miller, "From Edwards to Emerson," *Errand into the Wilderness*, Cambridge, Mass., 1956, pp. 184–204; George H. Williams, *Wilderness and Paradise in Christian Thought*, New York, 1962.

5. *Works*, I, 16; see *Journals*, V (1839), 310–11.

6. *Works*, I, 343–72; all quotations that follow are from this revised version of the lecture Emerson had read before the Mercantile Library Association of Boston on February 7, 1844, and published in the final number of the *Dial*, April, 1844.

7. James, *The American*, The New York Edition, 1907, ch. II.

8. Hart Crane, "Modern Poetry," reprinted in *The Collected Poems of Hart Crane*, ed. Waldo Frank, New York, 1946, p. 177; G. Ferris Cronkhite, "The Transcendental Railroad," *NEQ*, XXIV (1951), 306–28; Paul Ginestier, *The Poet and the Machine*, trans. Martin B. Friedman, Chapel Hill, 1961; Peter Viereck, "The Poet in the Machine Age," *JHI*, X (1949), 88–103; Jeremy Warburg, "Poetry and Industrialism: Some Refractory Material in Nineteenth-Century and Later English Verse," *MLR*, LIII (1958), 160–70; Emerson, *Works*, II, 325–43; *Works*, III, 9–45; see esp. II, 342–3, III, 23.

9. The Jacobean writer is unknown: Howard Schultz, "A Fragment of Jacobean Song in Thoreau's *Walden*," *MLN*, LXIII (1948), 271–2; Thoreau's itemized budget in "Economy."

10. Thoreau on wildness, "Spring"; Mircea Eliade, *Cosmos and History, The Myth of the Eternal Return*, trans. Willard R. Trask, New York, 1959, pp. 16–17; note: Renato Poggioli, "The Oaten Flute," *Harvard Library Bulletin*, XI (Spring, 1957), p. 152.

11. Carlyle, "Signs of the Times," see above, pp. 170ff; Thoreau, "Economy"; for Eli Terry and the clockmaking industry, see Dirk J. Struik, *Yankee Science in the Making*, p. 147; for mechanization and concepts of time, see S. Giedion, *Mechanization Takes Command, A Contribution to Anonymous History*, New York, 1948, pp. 77ff; Hans Meyerhoff, *Time in Literature*, Berkeley, 1960, esp. pp. 86–119; Lewis Mumford, *Technics and Civilization*, New York, 1934.

12. For Thoreau's failure as a farmer, see Leo Stoller, "Thoreau's Doctrine of Simplicity," *NEQ*, XXIX (Dec., 1956), 443–61; the ice crews, "The Pond in Winter"; "The Ponds."

13. "The Ponds," "Brute Neighbors," "The Pond in Winter"; for Thoreau's attitude toward "fact" and "truth," see Perry Miller, "Thoreau in the Context of International Romanticism," *NEQ*, XXXIV (June,

1961), 147–59 and Sherman Paul, *The Shores of America, Thoreau's Inward Exploration*, Urbana, 1958.

14. Hawthorne, *The American Notebooks*, ed. Randall Stewart, New Haven, 1932, p. 106.

15. *American Notebooks*, pp. 34–5.

16. *American Notebooks*, p. 42; for an earlier interpretation of "Ethan Brand" in relation to the Berkshire notes, see Leo Marx, "The Machine in the Garden," *NEQ*, XXIX (Mar., 1956), 27–42.

17. *Paradise Lost*, IX, 391–2, and X, 1060–93, *The Poems of John Milton*, ed. James Holly Hanford, New York, 1953; the fable of Prometheus, and especially the contrast between Prometheus and Epimetheus, offers a striking analogue to the pastoral design. Epimetheus, whose name implies a concern with the past, was committed to tradition, to a passive and almost feminine celebration of organicism. But Prometheus, who stole fire from the gods, was oriented toward the future and a technology in the service of man. His attitude was aggressively humanistic; Epimetheus was reverent toward the gods. The modern interest in the Promethean myth must be understood in relation to the development of technology in general and fire in particular. The most striking parallel is the contrast between Ahab and Ishmael in *Moby-Dick;* Ahab is a perverted or, in Richard Chase's phrase, a "sick Prometheus"; Richard Chase, *Herman Melville*, New York, 1949, and M. O. Percival, *A Reading of Moby-Dick*, Chicago, 1950. See also Gaston Bachelard, *The Psychoanalysis of Fire*, trans. Alan C. M. Ross, Boston, 1964.

18. Francis D. Klingender discusses John Martin's influence in *Art and the Industrial Revolution*, London, 1947, pp. 103–8, 192–6; Cole's "plagiarism" is described in Virgil Barker, *American Painting: History and Interpretation*, New York, 1950, p. 427; for Hawthorne's imagery, see Leo Marx, "The Machine in the Garden," *NEQ*, p. 37n.

19. Herman Melville, "Hawthorne and His Mosses," first published in *The Literary World*, 1850, *The Portable Melville*, ed. Jay Leyda, New York, 1952, p. 418; for a discussion of the cultural climate evoked by Hawthorne's sentimental pastoralism, see E. Douglas Branch, *The Sentimental Years, 1836–1860*, New York, 1934, and Herbert R. Brown, *The Sentimental Novel in America, 1789–1860*, Durham, N. C., 1940.

20. Quoted by Mark Van Doren, *Nathaniel Hawthorne*, American Men of Letters Series, New York, 1949, p. 138.

21. Melville to Hawthorne, June 1 (?), 1851, *The Letters of Herman Melville*, eds. Merrell R. Davis and William H. Gilman, New Haven, 1960, pp. 126–31. Copyright © 1960 by Yale University Press and used by permission of the publisher.

22. Emerson, *Works*, I, 35–6; George Orwell, "Politics and the English

Language," *Shooting an Elephant*, New York, 1950; Sigmund Freud, *Civilization and Its Discontents*, London, 1957, pp. 8f.

23. *Typee*, ch. 26; "Hawthorne and His Mosses," p. 413.

24. Tommo "unmanned," ch. 32; the psychological implications of Melville's symbolism have been discussed by a number of critics; I am particularly indebted to Henry A. Murray, "In Nomine Diaboli," *NEQ*, XXIV (Dec., 1951), 435–52, but I think Murray pays insufficient attention to Ishmael's rôle; see Newton Arvin, *Herman Melville*, American Men of Letters Series, New York, 1950, for the fullest Freudian interpretation.

25. *White-Jacket*, ch. 89; for Melville's reading in Carlyle, see Merton B. Sealts, "Melville's Reading, A Check-List of Books Owned and Borrowed, III," *Harvard Library Bulletin*, III (1949), 119–30, items 121–3.

26. *Moby-Dick*, chs. 10, 1, 41.

27. *Moby-Dick*, chs. 36, 35.

28. Emerson, *Works*, I, 16.

29. Ch. 41; ch. 42.

30. Chs. 48, 135.

31. The "complete man," ch. 108; ch. 41.

32. Chs. 55, 86.

33. Ch. 10.

34. Ch. 114.

35. Chs. 134, 135, 136.

36. Clemens to Howells, Oct. 24, 1874, *Mark Twain–Howells Letters, The Correspondence of Samuel L. Clemens and William D. Howells, 1872–1910*, eds. Henry Nash Smith and William M. Gibson, Cambridge, Mass., 1960, I, 34.

37. For the chronology of composition I follow Walter Blair, "When Was Huckleberry Finn Written?" *AL*, XXX (March, 1958), 1–25, and *Mark Twain and Huck Finn*, Berkeley, 1960.

38. *Life on the Mississippi*, ch. 9.

39. *Innocents*, ch. 22; *Roughing It*, ch. 23; *Innocents*, ch. 10; *Roughing It*, ch. 17; this is an abbreviated version of Leo Marx, "The Pilot and the Passenger: Landscape Conventions and the Style of *Huckleberry Finn*," *AL*, XXVIII (May, 1956), 129–46; for a somewhat different interpretation, see Roger B. Salomon, *Twain and the Image of History*, New Haven, 1961, esp. pp. 160–66.

40. Clemens to Howells, July 5, 1875, *Mark Twain–Howells Letters*, I, 91.

41. Ch. 1; this reading of *Huckleberry Finn* draws upon Henry Nash Smith, *Mark Twain, The Development of a Writer*, Cambridge, Mass., 1962.

42. Ch. 9.

43. Chs. 12, 18, 19.

44. Smith, *Mark Twain*, ch. 6; *A Tramp Abroad*, ch. 14.

45. Ch. 16.

46. Ch. 32.

47. Hemingway, *Green Hills of Africa*, New York, 1935, ch. 1; for a more complete statement of the case against the ending, see my essay, "Mr. Eliot, Mr. Trilling, and *Huckleberry Finn*," *American Scholar*, XXII (Autumn, 1953), 423–40.

48. Tony Tanner, "The Lost America — The Despair of Henry Adams and Mark Twain," *Modern Age*, V (Summer, 1961), 299–310, reprinted in *Mark Twain, A Collection of Critical Essays*, ed. Henry Nash Smith, Englewood Cliffs, N. J., 1963.

49. Trilling, *The Liberal Imagination, Essays on Literature and Society*, New York, 1950; Chase, *The American Novel and Its Tradition*, New York, 1957; Lewis, *The American Adam*.

50. Ch. 1; Donald Pizer has criticized this interpretation of Norris in "Synthetic Criticism and Frank Norris; Or, Mr. Marx, Mr. Taylor, and *The Octopus*," *AL*, XXXIV (Jan., 1963), 532–41.

51. Henry to Brooks Adams, June 18, 1903, quoted in J. C. Levenson, *The Mind and Art of Henry Adams*, Boston, 1957, p. 300.

52. Ch. 5.

53. Ch. 25.

54. Henry to Charles F. Adams, Jr., quoted in Ernest Samuels, *The Young Henry Adams*, Cambridge, Mass., 1948, p. 130.

55. James, *The American Scene*, ed. W. H. Auden, New York, 1946, pp. 13–18.

56. *American Scene*, pp. 38–44.

CHAPTER VI

1. Adams, *The Education of Henry Adams*, ch. 34. Quoted by permission of the Houghton Mifflin Company.

2. The Buchanan's lawn, ch. 1; the valley of ashes, ch. 2. The quotations from *The Great Gatsby* are by permission of Charles Scribner's Sons.

3. Ch. 4; ch. 3.

4. Ch. 6.

5. Ch. 1; ch. 2; ch. 9.

6. Ch. 6.

AFTERWORD

1. Leo Marx, "Hawthorne and Emerson: Studies in the Impact of the Machine Technology on the American Writer," Ph.D. diss., Harvard University, 1950. The degree is in the "History of American Civilization," or what is less grandly called, at most universities, "American studies."

2. So far as I can recall, Mason did not take exception to the point, nor did Joseph Schumpeter, who delivered a guest lecture on the "creative destruction" characteristic of unconstrained market economies.

3. Edmund Wilson, review of *The Last of the Provincials*, by Maxwell Geismer. *The New Yorker*, 23 (1947), 57.

4. I am thinking of abstract terms like *industrialization, industrial capitalism, mechanization, urbanization, rationalization, modernization, technological innovation*, etc.; here I want to acknowledge a large debt to Karl Polanyi, who made an invaluable contribution to my project during a week he spent with our research seminar at the University of Minnesota in 1953–54, and for his extraordinary book, *The Great Transformation*, New York: Rinehart, 1947.

5. In elucidating this body of images, I found the iconological method developed by Erwin Panofsky immensely suggestive. See *Studies in Iconology; Humanistic Themes in the Art of the Renaissance*, New York: Oxford University Press, 1939. I also benefited from the example of Francis D. Klingender, *Art and the Industrial Revolution*, London: Royle Publications, 1947.

6. I later included a version of this analysis in *The Machine in the Garden*, p.265ff; for the link to *Moby-Dick* see p. 308.

7. Brian Attebery has written an essay about theoretical issues that figure in my correspondence with Smith: "American Studies: A Not So Unscientific Method," *American Quarterly* 48 (1996), 316–43.

8. I finally categorized and compressed my examples of this rhetoric—I had collected enough to fill a small monograph—in a twenty-page segment (pp. 190–209), followed by a more thorough analysis (pp. 209–20) of Daniel Webster's strikingly ambivalent use of the rhetoric.

9. For the concept of "root metaphor" I was indebted to Stephen C. Pepper, *World Hypotheses, A Study in Evidence*, Berkeley: University of California Press, 1942.

10. It first appeared in the *New England Quarterly*, and is reprinted in Leo Marx, *The Pilot and the Passenger: Essays on Literature, Technology, and Culture in the United States*, New York: Oxford University Press, 1988, pp. 113–26.

11. In the opening line of the pastoral poem, "Directive."

12. In the Preface to his collected poems, "The Figure a Poem Makes."

13. In the opening chapter, "Economy," of *Walden*.

14. See above, p.116ff. For a recent elaboration of this argument see Leo Marx, "The Domination of Nature and the Redefinition of Progress," in *Progress: Fact or Illusion?* eds. Leo Marx and Bruce Mazlish, Ann Arbor: University of Michigan Press, 1998, pp. 201–18.

15. "Myth and Symbol in American Studies," *American Quarterly*, 27 (1972), 435–50.

16. Gene Wise, in noting the intricate personal and professional relationships among the "myth and symbol" writers, wrote: "Few political conspiracies have ever been so tightly interwoven as this one." "Paradigm Dreams in American Studies: A Cultural and Institutional History of the Movement," *American Quarterly*, 31 (1979), 311.

17. I later returned to the question of the political import of American pastoralism, including the role of the New Left, in "Pastoralism in America," in *Ideology and Classic American Literature*, eds. Sacvan Bercovitch and Myra Jehlen, New York: Cambridge University Press, 1986, pp. 36–69; and in "Does Pastoralism Have a Future?" in *The Pastoral Landscape*, ed. John Dixon Hunt, Washington, DC: National Gallery of Art, 1992, pp. 209–21.

18. Alain Touraine, *The Voice and the Eye: An Analysis of Social Movements*, Cambridge: Cambridge University Press, 1981.

Acknowledgments

MY WAY OF THINKING about literature and society was formed, years ago, when I was a student working with F. O. Matthiessen and Perry Miller. I have made no attempt to specify my debts to them. Anyone who knew them, or who knows their work, will have encountered their quite different influences everywhere in this book. It is only slightly easier for me to explain what I owe to my teacher and former colleague, Henry Nash Smith. For almost twenty years he has been trying, with endless patience and generosity, to clarify my ideas. He and John William Ward have read the entire manuscript, parts of it in several versions; their comments and their encouragement have made an inestimable difference to me. To Benjamin DeMott I owe a special thanks for his effort to prevent the writing of another book that is merely about books. Certain of my basic ideas were developed during the year I participated, with J. William Buchta, Arnold Rose, and Philip Wiener, in the Faculty Research Seminar in American Studies at the University of Minnesota under the resourceful chairmanship of Bernard Bowron.

Because I have been writing this book for a long time, and because parts of it appeared as separate essays, I have had the advice and criticism of more people than I have space to acknowledge. Samuel Holt Monk has given me the benefit of his expert knowledge of the eighteenth century and of his learning generally. C. L. Barber helped me to define the concept of pastoralism; he, Joseph Holmes Summers, and the late Newton Arvin contributed sound criticism of the Shakespeare chapter. To my colleagues George Kateb and George Rogers Taylor I am grateful for astute comments on large sections of the manuscript. It is impossible to give individual recognition to all of the students from whom I have learned in the doctoral seminar in American Studies at the University of Minnesota or in English 65 at Amherst College. Sheldon Meyer has given consistently wise editorial advice. Mary Ollmann, Joan Weston, and Evelyn Cooley have done all that they could to help me avoid confusion and error. Porter Dickinson and the staff of the Amherst College Library have been unfailingly considerate. Jane Marx and

Stephen Marx graciously helped with the final preparation of the manuscript.

Much of this book was written during a year of freedom from teaching duties made possible by generous grants from the Guggenheim Foundation and the Trustees of Amherst College.

For quotations from copyrighted material I have been given permission by: Ohio State University Press for the Randall Stewart edition of Nathaniel Hawthorne's *American Notebooks;* Penguin Books, Inc., for E. V. Rieu's translation of Virgil's eclogues, *The Pastoral Poems;* University of North Carolina Press for Louis Wright's edition of Robert Beverley, *The History and Present State of Virginia,* and William Peden's edition of Thomas Jefferson, *Notes on Virginia;* Houghton Mifflin Company for D. H. Lawrence's letter to Amy Lowell from S. Foster Damon, *Amy Lowell, A Chronicle,* and for *The Education of Henry Adams;* The Macmillan Company, New York, Mrs. William Butler Yeats, Messrs. Macmillan & Company, London, and A. P. Watt and Son, London, for the poem from *The Tower* by W. B. Yeats; Random House for William Faulkner, "The Bear," from *Go Down, Moses and Other Stories,* copyright 1942 by Curtis Publishing Co., copyright 1942 by William Faulkner; Yale University Press for Herman Melville's letter to Nathaniel Hawthorne from *The Letters of Herman Melville,* edited by Merrell R. Davis and William H. Gilman; Charles Scribner's Sons for F. Scott Fitzgerald, *The Great Gatsby.* Permission to reproduce the aerial photograph of the Rock Island Railroad Yards has been granted by the Standard Oil Company of New Jersey; the National Gallery of Art has provided the reproduction of George Inness, "The Lackawanna Valley"; and the Museum of Modern Art, New York, has given permission to reproduce Charles Sheeler's "American Landscape." I also am grateful to the editors of the following journals for permission to reprint my own work: *The Massachusetts Review* for parts of "Two Kingdoms of Force" (Autumn, 1959), used in Chapters 1 and 5, and for a revised version of "Shakespeare's American Fable" (Autumn, 1960), copyright © 1959, 1960, The Massachusetts Review, Inc.; *American Quarterly* for some material from "Literature and Covert Culture" (Winter, 1957), written with Bernard Bowron and Arnold Rose, used in Chapter 4; *New England Quarterly* for the title and a revised version of "The Machine in the Garden" (March, 1956), used in the section of Chapter 5 dealing with Hawthorne; *American Literature* for some passages from "The Pilot and the Passenger: Landscape Conventions and the Style of *Huckleberry Finn*" (May, 1956), used in Chapter 5.

For their critical reading of a draft of the Afterword, I am grateful to Jill Kerr Conway, Gail Fenske, Kenneth Keniston, Andrew Marx, Lucy Marx, Stephen Marx, Merritt Roe Smith, John Staudenmaier, and Alan Trachtenberg. I owe a special thanks to Sheldon Meyer for some 40 years of his unfailing encouragement and editorial sense.